HOW TO USE A LASER ENGRAVER

A Comprehensive Guide from Setup to Advanced Techniques

James Roland

All rights reserved. No part of this publication may be reproduced in any form or by any means, including photocopying, recording, or any other electronic or mechanical methods without the prior written permission of the publisher except in the case of brief quotations embodied in reviews and certain other non-commercial uses permitted by copyrights law.

Copyright © James Roland, 2024.

TABLE OF CONTENTS

Chapter 1 8
Introduction to Laser Engraving 8
   1.1 What is Laser Engraving? 8
   1.2 Types of Laser Engravers 9
   1.3 Applications of Laser Engraving 10
   1.4 Advantages and Disadvantages of Laser Engraving 11
   1.5 Safety Considerations 13
   1.6 Basic Terminology 15
   1.7 Overview of the Course 16
Chapter 2 19
Setting Up Your Laser Engraver 19
   2.1 Unboxing and Assembly 19
   2.2 Connecting to Power and Computer 20
   2.3 Installing Software and Drivers 21
   2.4 Calibrating the Laser 21
   2.5 Performing Test Runs 23
   2.6 Troubleshooting Common Setup Issues 23
   2.7 Maintenance and Cleaning Tips 24
Chapter 3 27
Choosing and Preparing Materials for Laser Engraving 27

3.1 Types of Materials Suitable for Laser Engraving 27

3.2 Understanding Material Properties 29

3.3 Preparing Materials for Engraving 31

3.4 Cleaning and Finishing Engraved Materials.... 32

3.5 Tips for Specific Materials (Wood, Metal, Acrylic, etc.) 33

3.6 Where to Buy Materials 35

3.7 Safety Precautions for Different Materials 36

Chapter 4 39

Designing for Laser Engraving 39

4.1 Choosing Design Software 39

4.2 Basic Design Principles 41

4.3 Creating Vector and Raster Images 42

4.4 Converting Images for Laser Engraving 43

4.5 Importing and Exporting Files 44

4.6 Tips for Effective Designs 45

4.7 Common Design Mistakes to Avoid 46

CHAPTER 5 48

LASER ENGRAVING BASICS 48

5.1 Understanding Laser Settings (Power, Speed, DPI) 48

5.2 Choosing the Right Settings for Different Materials 50

5.3 Engraving Text 51

5.4 Engraving Images.. 53

5.5 Creating Depth and Dimension 54

5.6 Testing and Adjusting Settings 55

5.7 Troubleshooting Common Engraving Issues... 56

CHAPTER 6 59

ADVANCED LASER ENGRAVING TECHNIQUES 59

6.1 Engraving on Curved Surfaces........................... 59

6.2 Creating 3D Engravings 61

6.3 Combining Engraving and Cutting................... 62

6.4 Photo Engraving... 62

6.5 Engraving with Multiple Colors 63

6.6 Using Fill Patterns and Textures........................ 64

6.7 Experimenting with Different Techniques 65

CHAPTER 7 67

LASER CUTTING 67

7.1 Understanding Laser Cutting Settings.............. 67

7.2 Choosing the Right Settings for Different Materials 69

7.3 Cutting Basic Shapes... 71

7.4 Cutting Complex Designs 72

7.5 Cutting through Thick Materials 74

7.6 Tips for Precise Cutting....................................... 75

7.7 Troubleshooting Common Cutting Issues......... 77

CHAPTER 8 80

5

CREATING PROJECTS 80
  8.1 Brainstorming Project Ideas 80
  8.2 Designing and Planning Your Projects 82
  8.3 Gathering Materials and Tools 83
  8.4 Executing Your Projects Step-by-Step 84
  8.5 Finishing and Polishing Your Projects 86
  8.6 Showcasing Your Creations 87
  8.7 Finding Inspiration Online 88
CHAPTER 9 92
MAINTAINING YOUR LASER ENGRAVER 92
  9.1 Regular Cleaning and Maintenance 92
  9.2 Replacing Parts and Consumables 94
  9.3 Troubleshooting Common Problems 95
  9.4 Extending the Lifespan of Your Engraver 97
  9.5 Safety Checks and Inspections 98
  9.6 Calibrating and Aligning the Laser 99
  9.7 Professional Servicing and Repairs 100
CHAPTER 10 102
ADVANCED SOFTWARE AND DESIGN 102
  10.1 Exploring Advanced Design Software 102
  10.2 Creating Complex Designs 104
  10.3 Using CAD Software for Precision 106
  10.4 Incorporating 3D Modeling 107
  10.5 Generating G-Code for Engraving 108

10.6 Optimizing Designs for Efficiency 109

10.7 Troubleshooting Software Issues 110

CHAPTER 11 113

BUSINESS AND ENTREPRENEURSHIP WITH LASER ENGRAVING 113

11.1 Starting a Laser Engraving Business 114

11.2 Marketing and Promoting Your Services 116

11.3 Pricing Your Products and Services 118

11.4 Managing Orders and Production 119

11.5 Building a Customer Base 121

11.6 Expanding Your Business 122

11.7 Legal and Regulatory Considerations 123

CHAPTER 12 126

THE FUTURE OF LASER ENGRAVING 126

12.1 Emerging Trends and Technologies 126

12.2 New Materials and Applications 128

12.3 Innovations in Laser Engraving Software 130

12.4 Environmental Impact and Sustainability ... 131

12.5 The Role of AI in Laser Engraving 132

12.6 Career Opportunities in the Field 134

CHAPTER 1

INTRODUCTION TO LASER ENGRAVING

Laser engraving is a fascinating and versatile technology that has revolutionized the way we create, personalize, and customize objects. From intricate designs on jewelry to industrial markings on machine parts, laser engraving offers precision, speed, and endless possibilities. In this chapter, we'll delve into the fundamentals of laser engraving, exploring its history, technology, applications, advantages, disadvantages, safety considerations, and essential terminology. Whether you're a curious beginner or a seasoned professional, this chapter will provide a solid foundation for your journey into the world of laser engraving.

1.1 What is Laser Engraving?

At its core, laser engraving is a subtractive manufacturing process that uses a focused laser beam to remove material from a surface, creating a permanent mark or design. The laser beam, typically generated by a CO_2 or fiber laser, is directed onto the material's surface, where it interacts with the molecules, causing them to vaporize or melt. This controlled removal of material results in precise and detailed engravings.

The process is highly versatile, allowing for engraving on a wide range of materials, including wood, metal, glass, acrylic, leather, and even food. The depth and intensity of the engraving can be controlled by adjusting the laser's power, speed, and focus, enabling the creation of both subtle markings and deep reliefs.

1.2 Types of Laser Engravers

There are various types of laser engravers available, each with its own strengths and applications. Let's take a closer look at the most common types:

- **CO2 Laser Engravers:** These are the most widely used type of laser engravers, known for their versatility and ability to engrave on a wide range of materials. CO2 lasers emit infrared light, which is readily absorbed by many non-metals, making them ideal for engraving wood, acrylic, leather, fabric, and other organic materials.
- **Fiber Laser Engravers:** Fiber lasers emit a highly focused beam of light, making them perfect for engraving on metal surfaces. They are known for their precision and speed, making them a popular choice for industrial marking, barcode engraving, and other high-precision applications.
- **YAG Laser Engravers:** YAG (yttrium aluminum garnet) lasers are similar to fiber lasers but operate at a different wavelength. They are often used for engraving on metals

and certain plastics, and are known for their ability to create deep and permanent marks.
- **UV Laser Engravers:** UV (ultraviolet) lasers emit a shorter wavelength of light compared to other types of lasers. This allows them to engrave on a wider range of materials, including glass, ceramics, and some plastics, with minimal heat damage.

When choosing a laser engraver, it's important to consider the types of materials you'll be working with, the desired engraving quality, and your budget.

1.3 Applications of Laser Engraving

Laser engraving has found applications in a multitude of industries and creative endeavors. Here are just a few examples:

- **Industrial Marking:** Laser engraving is widely used in industrial settings to mark machine parts, tools, and other products with serial numbers, barcodes, logos, and other identifying information. The permanent and tamper-proof nature of laser engraving makes it ideal for traceability and quality control.
- **Personalization and Customization:** Laser engraving allows for the creation of unique and personalized items, such as jewelry, trophies, awards, gifts, and promotional products. It's a popular choice for adding names, dates, messages, and designs to make items truly special.

- **Artistic Expression:** Artists and designers use laser engraving to create intricate artwork, sculptures, and decorative pieces. The ability to engrave on various materials with high precision opens up endless creative possibilities.
- **Medical Devices:** Laser engraving is used to mark medical instruments and implants with unique identifiers, ensuring traceability and safety.
- **Electronics Manufacturing:** Laser engraving is employed in the electronics industry to mark circuit boards, components, and casings with part numbers, logos, and other information.
- **Aerospace Industry:** Laser engraving is used to mark aerospace components with critical identification and traceability information.

These are just a few examples of the diverse applications of laser engraving. As technology continues to advance, we can expect even more innovative uses for this versatile process.

1.4 Advantages and Disadvantages of Laser Engraving

Laser engraving offers several advantages over traditional engraving methods:

- **Precision and Accuracy:** Laser engraving allows for incredibly precise and detailed markings, even on complex designs and small surfaces. The laser beam can be focused to a

tiny spot, enabling the creation of intricate patterns and fine lines.
- **Speed and Efficiency:** Laser engraving is a fast process, especially compared to manual engraving methods. The laser beam can quickly remove material, making it suitable for high-volume production and rapid prototyping.
- **Versatility:** Laser engravers can work with a wide variety of materials, including metals, plastics, wood, glass, leather, and more. This versatility opens up a vast range of creative and industrial applications.
- **Non-Contact Process:** Laser engraving is a non-contact process, meaning there's no physical pressure applied to the material. This eliminates the risk of damage to delicate or fragile items.
- **Permanent Marking:** Laser engravings are permanent and resistant to wear and tear, making them ideal for applications where durability and longevity are essential.
- **Automation and Repeatability:** Laser engraving can be easily automated, allowing for consistent and repeatable results. This is crucial for industrial applications where high precision and consistency are required.

However, laser engraving also has some disadvantages:

- **Cost:** Laser engravers can be expensive, especially high-powered models with

advanced features. The initial investment can be a barrier for some individuals and businesses.
- **Material Limitations:** While laser engravers can work with a wide range of materials, some materials are not suitable for laser engraving due to their composition or properties. For example, certain plastics may melt or release harmful fumes when exposed to laser beams.
- **Safety Hazards:** Laser engraving involves the use of powerful laser beams, which can be hazardous to the eyes and skin. Proper safety precautions, such as wearing protective eyewear and operating the engraver in a well-ventilated area, are essential.
- **Learning Curve:** Mastering laser engraving techniques and software can take time and practice. It's important to invest in training and resources to ensure safe and effective operation.

Overall, the advantages of laser engraving often outweigh the disadvantages, making it a valuable tool for a wide range of applications.

1.5 Safety Considerations

Laser engraving involves the use of powerful laser beams that can pose risks to human health if not handled properly. It's crucial to prioritize safety when working with laser engravers. Here are some essential safety considerations:

- **Eye Protection:** Always wear appropriate laser safety eyewear that is specifically designed for the wavelength of the laser you're using. Laser beams can cause serious eye damage, including blindness, if they enter the eye directly.
- **Skin Protection:** Avoid direct contact with the laser beam, as it can cause burns. Wear gloves and long sleeves when operating the engraver, and avoid pointing the laser at yourself or others.
- **Ventilation:** Laser engraving can produce fumes and particulate matter, depending on the material being engraved. Always operate the engraver in a well-ventilated area or use a fume extraction system to remove harmful emissions.
- **Fire Hazards:** Some materials can ignite when exposed to laser beams. Keep flammable materials away from the engraving area and have a fire extinguisher readily available.
- **Electrical Safety:** Follow electrical safety guidelines when connecting and operating the laser engraver. Avoid using damaged cables or operating the engraver in wet or damp conditions.
- **Training and Certification:** If you're using a laser engraver in a professional setting, ensure that you and your staff receive proper training and certification in laser safety.

By following these safety precautions and adhering to best practices, you can minimize the risks associated with laser engraving and create a safe working environment.

1.6 Basic Terminology

Before we dive deeper into laser engraving, let's familiarize ourselves with some basic terminology:

- **Laser:** A device that emits a narrow beam of light with a single wavelength.
- **Wavelength:** The distance between two consecutive peaks of a wave, typically measured in nanometers (nm).
- **Power:** The amount of energy emitted by the laser beam, measured in watts (W).
- **Speed:** The rate at which the laser beam moves across the material, measured in millimeters per second (mm/s).
- **DPI (Dots Per Inch):** The resolution of the engraving, which determines the level of detail.
- **Vector Image:** An image composed of lines and curves defined by mathematical equations.
- **Raster Image:** An image composed of pixels, similar to a photograph.
- **Focus:** The point at which the laser beam is most concentrated and produces the most intense heat.
- **Bed:** The surface on which the material is placed for engraving.
- **Enclosure:** The protective housing that surrounds the laser engraver.

- **Ventilation:** The process of removing fumes and particulate matter generated during engraving.
- **Safety Eyewear:** Special glasses that protect the eyes from laser radiation.

Understanding these terms will help you communicate effectively with other laser engraving enthusiasts and professionals, and it will make it easier to follow tutorials and instructions.

1.7 Overview of the Course

In this comprehensive course on laser engraving, we'll cover a wide range of topics, from basic setup and operation to advanced techniques and creative applications. We'll guide you through the entire process, step-by-step, so you can confidently use a laser engraver to bring your ideas to life.

Here's a brief overview of what you can expect to learn:

How to choose the right laser engraver for your needs and budget

- How to set up and calibrate your laser engraver
- How to choose and prepare materials for engraving
- How to design for laser engraving using various software

- How to master basic and advanced laser engraving techniques
- How to create stunning projects, from personalized gifts to intricate artwork
- How to maintain your laser engraver for optimal performance and longevity
- How to explore advanced software and design techniques
- How to start and grow a laser engraving business

We'll also share tips, tricks, and troubleshooting advice to help you overcome common challenges and achieve professional-quality results. By the end of this course, you'll have the knowledge and skills to unleash your creativity and make the most of this powerful technology.

Laser engraving is a captivating technology that offers endless possibilities for creativity and innovation. Whether you're interested in personalizing gifts, creating custom products, or marking industrial components, laser engraving provides a versatile and efficient solution.

In this chapter, we've laid the groundwork for your laser engraving journey. We've explored the fundamentals of the technology, its applications, advantages, disadvantages, safety considerations, and essential terminology. Armed with this knowledge, you're ready to dive deeper into the exciting world of laser engraving.

In the following chapters, we'll guide you through the entire process, step by step, from setting up your engraver to mastering advanced techniques. Get ready to unleash your creativity and transform your ideas into reality with the power of laser engraving!

CHAPTER 2

SETTING UP YOUR LASER ENGRAVER

Congratulations on taking the next step in your laser engraving journey! Now that you have a basic understanding of what laser engraving is and its various applications, it's time to roll up your sleeves and get your hands on the machine. This chapter is your comprehensive guide to setting up your laser engraver, ensuring it's ready to bring your creative visions to life. We'll cover everything from unboxing and assembly to connecting it to your computer and performing those crucial test runs. Don't worry if you're a beginner; we'll take it slow and steady, explaining each step in detail so you can confidently embark on your engraving adventures.

2.1 Unboxing and Assembly

The moment you've been waiting for has finally arrived—your laser engraver has arrived! The unboxing process is exciting, but it's important to approach it with care. Start by carefully inspecting the packaging for any signs of damage. Once you're sure everything is intact, it's time to unveil your new tool.

Carefully remove the laser engraver from the packaging, paying attention to any loose parts or accessories. Most laser engravers come partially assembled, so you'll need to follow the manufacturer's instructions to complete the setup. This typically involves attaching components like the laser head, control panel, and exhaust fan.

Take your time during assembly, ensuring that all parts are securely fastened and aligned properly. Refer to the provided diagrams and instructions for guidance. If you encounter any difficulties, don't hesitate to contact the manufacturer's customer support for assistance.

2.2 Connecting to Power and Computer

Once your laser engraver is assembled, the next step is to connect it to a power source and your computer. Ensure that you have a compatible power outlet and a USB cable to establish the connection.

Begin by plugging the power cord into the laser engraver and then into the wall outlet. Turn on the engraver's power switch, and you should see lights or indicators on the control panel illuminating, indicating that the machine is receiving power.

Next, connect the USB cable from the laser engraver to your computer. Depending on your engraver's model and software, the computer may automatically detect the device and install the necessary drivers. If not, you'll need to manually install the drivers provided by the manufacturer. Refer to the user manual for detailed instructions on driver installation.

2.3 Installing Software and Drivers

Laser engravers require specialized software to control the engraving process and send designs to the machine. Most manufacturers provide their own software, which is often included with the engraver or available for download from their website.

Installing the software is usually a straightforward process. Follow the on-screen instructions provided by the installer. If you encounter any issues, consult the software's documentation or contact the manufacturer's support team for assistance.

In addition to the main software, you may also need to install drivers specific to your laser engraver model. These drivers facilitate communication between the software and the engraver, ensuring smooth and accurate operation. The drivers are typically included with the software or available on the manufacturer's website.

Once the software and drivers are installed, you'll be ready to start designing and sending your creations to the laser engraver.

2.4 Calibrating the Laser

Calibration is a crucial step in setting up your laser engraver to ensure accurate and consistent results. Calibration involves adjusting the laser's focus, power, and speed settings to match the specific materials you'll be engraving.

Start by placing a piece of scrap material on the engraver's bed. This will serve as your test subject for calibration. Open the laser engraver software and locate the calibration settings. These settings may vary depending on your software and engraver model, but they typically include options for adjusting focus, power, and speed.

Begin by adjusting the focus of the laser beam. The focus determines how concentrated the beam is at the point where it hits the material. A well-focused beam will produce sharper and more precise engravings. Refer to your engraver's manual for instructions on how to adjust the focus.

Next, adjust the laser's power setting. The power determines the intensity of the laser beam, which affects the depth and darkness of the engraving. Start with a low power setting and gradually increase it until you achieve the desired result on your test material.

Finally, adjust the speed setting. The speed determines how quickly the laser beam moves across the material. A slower speed will result in deeper engravings, while a faster speed will produce shallower markings. Experiment with different speed settings to find the optimal balance for your materials and designs.

2.5 Performing Test Runs

Once you've calibrated the laser engraver, it's time to perform some test runs to ensure everything is working properly. Choose a simple design, such as a geometric shape or a few lines of text, and send it to the engraver.

Observe the engraving process closely, paying attention to any unusual noises, vibrations, or smoke. If you notice any issues, stop the engraver immediately and investigate the cause.

If the test runs are successful and the engravings look as expected, congratulations! Your laser engraver is ready for action.

2.6 Troubleshooting Common Setup Issues

Even with careful setup and calibration, you may encounter some common issues with your laser engraver. Here are a few troubleshooting tips for common problems:

- **The laser beam isn't firing:** Check the power connections and ensure that the laser engraver is turned on. Verify that the software is sending the correct commands to the engraver. If the problem persists, consult the user manual or contact the manufacturer's support.
- **The engravings are blurry or uneven:** Check the focus of the laser beam and adjust it as

needed. Ensure that the material is securely placed on the bed and that the engraver is level.
- **The engraver is making strange noises:** Check for loose parts or obstructions in the engraver's path. Clean the mirrors and lenses to remove any dust or debris.
- **The software isn't communicating with the engraver:** Verify that the USB cable is securely connected and that the correct drivers are installed. Restart both the software and the engraver.
- **The engravings are too light or too dark:** Adjust the power and speed settings as needed. Experiment with different settings on a scrap material until you achieve the desired results.

If you're unable to resolve the issue on your own, don't hesitate to seek help from online forums, user groups, or the manufacturer's support team.

2.7 Maintenance and Cleaning Tips

To ensure the longevity and optimal performance of your laser engraver, regular maintenance and cleaning are essential. Here are some tips to keep your engraver in top shape:

- **Clean the mirrors and lenses:** Dust and debris can accumulate on the mirrors and lenses, affecting the laser beam's quality and accuracy.

Clean them regularly using a soft, lint-free cloth and a specialized cleaning solution.
- **Clean the bed:** Remove any debris or residue from the bed after each use. A clean bed ensures accurate placement of materials and prevents damage to the engraver.
- **Check the exhaust system:** The exhaust fan and filter are crucial for removing fumes and particulate matter generated during engraving. Check them regularly and clean or replace them as needed.
- **Lubricate moving parts:** Some laser engravers have moving parts that require lubrication. Refer to the user manual for instructions on how to lubricate these parts properly.
- **Store the engraver properly:** When not in use, store the laser engraver in a clean, dry, and dust-free environment. Cover it with a protective cloth or case to prevent damage.

By following these maintenance and cleaning tips, you can extend the lifespan of your laser engraver and ensure it continues to deliver high-quality results.

Setting up your laser engraver is an exciting first step in your creative journey. By following the steps outlined in this chapter, you can ensure that your engraver is properly assembled, connected, calibrated, and ready to tackle any project you throw at it.

Remember, patience and attention to detail are key during the setup process. Take your time, follow the instructions carefully, and don't hesitate to seek help if you encounter any difficulties. With a little effort and practice, you'll soon be confidently creating stunning engravings and unleashing your artistic potential.

CHAPTER 3

CHOOSING AND PREPARING MATERIALS FOR LASER ENGRAVING

Welcome to the exciting world of materials! In this chapter, we'll dive into the vast array of materials that can be transformed through the magic of laser engraving. Choosing the right material is crucial for achieving the desired results in your projects. We'll explore various material types, their unique properties, how to prepare them for engraving, and the finishing touches that will elevate your creations. Whether you're aiming for a rustic charm with wood, a sleek finish with metal, or vibrant colors with acrylic, this chapter will equip you with the knowledge to make informed decisions and unleash your creativity with confidence.

3.1 Types of Materials Suitable for Laser Engraving

Laser engraving opens up a world of possibilities when it comes to materials. From organic substances like wood and leather to synthetic materials like acrylic and plastics, the laser's versatile beam can etch its mark on a diverse range. Let's explore some of the most popular materials for laser engraving:

- **Wood:** A classic choice for laser engraving, wood offers a natural warmth and character that is unmatched. Different wood species like birch, maple, walnut, and bamboo offer varying densities and grain patterns, allowing for diverse engraving effects. Wood is perfect for personalized gifts, signage, artwork, and decorative items.
- **Metal:** Laser engraving on metal creates a sophisticated and permanent mark. Metals like stainless steel, aluminum, brass, and copper are popular choices for industrial applications, jewelry, awards, and personalized tools. The engraving process on metal often involves oxidation, resulting in a contrasting color change that enhances the design's visibility.
- **Acrylic:** This versatile plastic material is a favorite among laser engraving enthusiasts. Acrylic comes in various colors and finishes, offering a vibrant palette for creative expression. It's perfect for creating signs, displays, key chains, phone cases, and other decorative items.
- **Leather:** Laser engraving on leather produces a classic and elegant look. The laser can create intricate patterns, logos, or text on leather goods like wallets, belts, notebooks, and even shoes. The natural texture of leather complements the laser's precision, resulting in beautiful and personalized creations.
- **Glass:** Laser engraving on glass creates a frosted or etched effect, adding a touch of sophistication to the material. It's commonly

used for creating glassware, awards, trophies, and decorative panels. With the right techniques, you can even achieve photorealistic engravings on glass.
- **Fabric:** Certain fabrics, like cotton and denim, can be laser engraved to create unique designs and patterns. This technique is often used for customizing apparel, home decor items, and promotional products.
- **Stone:** Laser engraving on stone, such as marble or granite, produces a durable and elegant mark. This technique is often used for creating memorials, tombstones, architectural signage, and decorative elements.
- **Ceramics:** Laser engraving on ceramics can create beautiful and intricate designs on tiles, mugs, plates, and other decorative items.

The list of materials suitable for laser engraving is constantly expanding as new materials and technologies emerge. Don't be afraid to experiment with different materials to discover new possibilities and unleash your creativity.

3.2 Understanding Material Properties

Each material has unique properties that affect how it interacts with the laser beam and how the engraving turns out. Understanding these properties is essential for choosing the right material for your project and achieving optimal results.

- **Density:** The density of a material determines how much laser energy it absorbs. Denser materials, like metals, require higher laser power settings, while less dense materials, like wood, can be engraved with lower power.
- **Reflectivity:** Some materials, like polished metals, are highly reflective and can bounce the laser beam away, making it difficult to engrave. In such cases, it may be necessary to apply a masking material or use a laser engraver specifically designed for reflective surfaces.
- **Thermal Conductivity:** The thermal conductivity of a material determines how quickly it dissipates heat. Materials with high thermal conductivity, like metals, can handle higher laser power without warping or melting, while materials with low thermal conductivity, like wood, may require slower speeds or multiple passes to prevent burning.
- **Surface Finish:** The surface finish of a material can affect the appearance of the engraving. Smooth surfaces generally produce cleaner and more precise engravings, while rough or textured surfaces may result in less defined markings.
- **Thickness:** The thickness of a material determines how deep the laser beam can penetrate. Thicker materials may require higher power settings or multiple passes to achieve the desired engraving depth.
- **Toxicity:** Some materials, like certain plastics and treated woods, may release harmful fumes

when exposed to laser beams. Always check the material safety data sheet (MSDS) before engraving any material to ensure it's safe to use with a laser.

By understanding the properties of different materials, you can choose the most suitable one for your project and adjust the laser settings accordingly to achieve optimal results.

3.3 Preparing Materials for Engraving

Proper preparation of materials is essential for achieving high-quality engravings. Here are some general steps to follow when preparing materials for laser engraving:

- **Cleaning:** Clean the surface of the material to remove any dust, dirt, or debris that could interfere with the engraving process. Use a soft cloth, compressed air, or a mild cleaning solution appropriate for the material.
- **Masking:** Apply masking tape or masking film to the surface of the material to protect it from scratches and burns caused by the laser beam. This is especially important for delicate materials like glass or certain plastics.
- **Securing:** Secure the material to the engraver's bed using clamps or tape to prevent it from shifting during engraving. This is particularly important for thin or lightweight materials.
- **Leveling:** Ensure that the material is level on the bed to ensure even engraving. Use shims or

leveling tools to adjust the material's height if necessary.
- **Focusing:** Set the correct focus distance for the laser beam based on the material's thickness and the desired engraving depth. This will ensure a sharp and precise engraving.
- **Testing:** Before engraving your final design, always perform a test run on a scrap piece of the same material to verify the settings and ensure the desired results.

The specific preparation steps may vary depending on the material you're using. For example, metal surfaces may require additional cleaning with solvents or abrasives to remove coatings or oxides. Always refer to the material manufacturer's recommendations and the laser engraver's user manual for specific preparation instructions.

3.4 Cleaning and Finishing Engraved Materials

After the engraving process is complete, it's important to clean and finish the engraved material to enhance its appearance and protect it from damage. Here are some common cleaning and finishing techniques:

- **Removing Masking:** Carefully remove any masking tape or film from the surface of the material. Use a gentle touch to avoid scratching or damaging the engraving.
- **Wiping:** Wipe the surface of the engraved material with a soft cloth or sponge to remove

any residue or debris. Use a mild cleaning solution if necessary, but avoid harsh chemicals that could damage the material or the engraving.
- **Polishing:** For metal engravings, polishing can enhance the contrast and shine of the engraving. Use a metal polishing cloth or paste to buff the surface until it shines.
- **Sealing:** Some materials, like wood and leather, may benefit from sealing to protect the engraving and enhance its longevity. Apply a clear sealant or finish appropriate for the material to protect it from moisture, dirt, and wear and tear.
- **Painting or Staining:** For materials like wood, you can paint or stain the engraved areas to add color and contrast. Use a fine-tipped brush or airbrush to apply the paint or stain carefully.

The specific cleaning and finishing techniques will depend on the material you're using and the desired effect. Always test any cleaning or finishing products on a scrap piece of material before applying them to your final project.

3.5 Tips for Specific Materials (Wood, Metal, Acrylic, etc.)

Each material requires specific considerations and techniques for laser engraving. Here are some tips for working with common materials:

- **Wood:** Choose woods with fine grain patterns and avoid knots or imperfections. Sand the surface smooth before engraving. Use masking tape to prevent charring and burning.
- **Metal:** Clean the surface thoroughly to remove any coatings or oils. Apply a masking material or use a laser engraver designed for metal to prevent reflections.
- **Acrylic:** Choose cast acrylic over extruded acrylic for better engraving results. Use masking tape to prevent chipping and cracking.
- **Leather:** Use a high-quality leather with a smooth surface. Clean the surface with a damp cloth and let it dry before engraving.
- **Glass:** Clean the surface with a glass cleaner and lint-free cloth. Apply masking tape to prevent chipping and shattering. Use a lower power setting and slower speed to avoid overheating.
- **Fabric:** Choose natural fabrics like cotton or linen. Prewash the fabric to remove any sizing or finishes. Use a low power setting and high speed to avoid burning.

Remember to always test your settings on a scrap piece of the same material before engraving your final project. This will help you avoid costly mistakes and ensure the best possible results.

3.6 Where to Buy Materials

You can find materials for laser engraving from various sources, including:

- **Online Retailers:** Online retailers like Amazon, Etsy, and specialty laser engraving suppliers offer a vast selection of materials in different sizes, thicknesses, and colors.
- **Local Craft Stores:** Craft stores often carry a variety of materials suitable for laser engraving, such as wood sheets, acrylic blanks, and leather pieces.

Home Improvement Stores: Home improvement stores like Home Depot and Lowe's may have materials like plywood, MDF (medium-density fiberboard), and metal sheets that can be used for laser engraving.

- **Specialty Suppliers:** There are numerous online and brick-and-mortar stores that specialize in laser engraving materials. These suppliers often offer a wider range of options, including pre-cut shapes, engraving blanks, and specialty materials like anodized aluminum or two-tone acrylic.
- **Scrap Yards and Salvage Stores:** For metal enthusiasts, scrap yards and salvage stores can be treasure troves of interesting materials for laser engraving. You might find old tools, machine parts, or unique metal pieces that can be repurposed into creative projects.

- **Thrift Stores and Flea Markets:** These places can be a great source for finding vintage items that can be personalized or upcycled with laser engraving. Look for old wooden boxes, metal trays, or glass objects that can be given a new lease on life with a custom engraving.

When choosing a supplier, consider factors like price, quality, shipping costs, and the variety of materials offered. It's also a good idea to read reviews and compare prices before making a purchase.

3.7 Safety Precautions for Different Materials

While laser engraving is a versatile and relatively safe process, it's important to be aware of the potential hazards associated with different materials. Some materials may release harmful fumes or particulate matter when exposed to laser beams, while others may catch fire or melt.

Here are some general safety precautions to keep in mind when working with different materials:

- **Always wear safety glasses:** Laser safety glasses are essential for protecting your eyes from harmful laser radiation. Choose glasses that are specifically designed for the wavelength of your laser engraver.
- **Work in a well-ventilated area:** Laser engraving can produce fumes and particulate matter, depending on the material being

engraved. Always work in a well-ventilated area or use a fume extraction system to remove harmful emissions.
- **Avoid engraving toxic materials:** Some materials, like certain plastics and treated woods, may release toxic fumes when engraved. Always check the material safety data sheet (MSDS) before engraving any material to ensure it's safe to use with a laser.
- **Keep flammable materials away:** Some materials, like paper, cardboard, and certain fabrics, can catch fire when exposed to laser beams. Keep these materials away from the engraving area and have a fire extinguisher readily available.
- **Wear gloves and protective clothing:** When handling materials, especially metals, wear gloves and protective clothing to prevent burns or cuts.
- **Clean up spills immediately:** Spills of any material, especially liquids, should be cleaned up immediately to prevent accidents.
- **Follow manufacturer's instructions:** Always follow the manufacturer's instructions for your laser engraver and the materials you're using. This will help ensure safe and optimal results.

Choosing and preparing materials for laser engraving is a crucial step in creating successful and satisfying projects. By understanding the properties of different materials, preparing them properly, and taking necessary safety precautions, you can unleash your creativity and achieve stunning results.

Remember, the world of materials is vast and constantly evolving. Don't be afraid to experiment with different materials, explore new techniques, and push the boundaries of your creativity. With the right knowledge and tools, the possibilities are endless.

Chapter 4

Designing for Laser Engraving

Welcome to the heart of creativity and precision! In this chapter, we'll embark on a journey into the world of designing for laser engraving. This is where your artistic vision and technical skills converge to create stunning and personalized works of art. We'll delve into the intricacies of choosing the right design software, understanding fundamental design principles, creating vector and raster images, converting images for laser engraving, importing and exporting files, and uncovering valuable tips for crafting effective designs. By the end of this chapter, you'll be equipped with the knowledge and tools to transform your ideas into captivating designs ready to be etched onto your chosen materials. So, let's ignite your creative spark and explore the endless possibilities of designing for laser engraving!

4.1 Choosing Design Software

The first step in your design journey is selecting the right software to bring your ideas to life. The market offers a wide range of design software, each with its own strengths and features. The ideal software for you will depend on your budget, skill level, and the complexity of your designs.

- **Inkscape (Free):** A powerful open-source vector graphics editor that is perfect for beginners and those on a budget. Inkscape offers a comprehensive set of tools for creating and editing vector images, making it a popular choice for laser engraving designs.
- **Adobe Illustrator (Paid):** A professional-grade vector graphics editor widely used by graphic designers and artists. Illustrator offers advanced features and a robust toolset for creating complex and intricate designs.
- **CorelDRAW (Paid):** Another popular professional vector graphics editor that provides a user-friendly interface and powerful design capabilities. CorelDRAW is a versatile choice for both graphic design and laser engraving projects.
- **LightBurn (Paid):** A dedicated laser engraving software that streamlines the design and engraving process. LightBurn offers features like image tracing, node editing, and laser optimization, making it a valuable tool for laser engraving enthusiasts.
- **RDWorks (Often Included with Laser Engravers):** Many laser engravers come bundled with RDWorks software, which provides basic design tools and control over the engraving process. While not as comprehensive as dedicated design software, RDWorks is a good starting point for beginners.

When choosing design software, consider factors like ease of use, features, compatibility with your laser engraver, and cost. It's also helpful to read reviews and compare different options before making a decision.

4.2 Basic Design Principles

Whether you're a seasoned designer or just starting, understanding basic design principles is essential for creating visually appealing and effective laser engraving designs. Here are some key principles to keep in mind:

- **Simplicity:** Keep your designs simple and easy to read. Avoid cluttering the design with too many elements or intricate details that might get lost in the engraving process.
- **Contrast:** Create contrast between the engraved areas and the background to make your design stand out. Use contrasting colors, textures, or depths to enhance visibility.
- **Balance:** Strive for a balanced composition that distributes visual weight evenly throughout the design. This can be achieved through symmetry, asymmetry, or the use of negative space.
- **Alignment:** Align elements within your design to create a sense of order and coherence. Use grids, guides, and snapping tools to ensure precise alignment.
- **Typography:** Choose fonts that are legible and appropriate for the purpose of the engraving.

Avoid using too many different fonts or overly decorative fonts that might be difficult to read.
- **Spacing:** Pay attention to the spacing between letters, words, and elements within your design. Proper spacing enhances readability and visual appeal.
- **Hierarchy:** Establish a clear visual hierarchy within your design to guide the viewer's eye. Use size, color, and contrast to emphasize important elements and create a focal point.

By applying these basic design principles, you can create laser engraving designs that are not only visually appealing but also effective in conveying your message or artistic vision.

4.3 Creating Vector and Raster Images

Laser engraving works with two primary types of images: vector and raster. Understanding the difference between these two image types is crucial for achieving optimal results.

- **Vector Images:** Vector images are composed of lines and curves defined by mathematical equations. They can be scaled up or down without losing quality, making them ideal for laser engraving. Vector images are typically created using vector graphics editors like Inkscape, Adobe Illustrator, or CorelDRAW.
- **Raster Images:** Raster images are composed of pixels, similar to a photograph. They have a fixed resolution and can become blurry or

pixelated when scaled up. While raster images can be used for laser engraving, they often require conversion to vector format to ensure sharp and precise results.

When creating designs for laser engraving, it's generally recommended to use vector images whenever possible. Vector images offer greater flexibility, scalability, and control over the engraving process. However, if you're working with photographs or other raster images, there are ways to convert them to vector format using tracing tools or specialized software.

4.4 Converting Images for Laser Engraving

Converting raster images to vector format is a common practice in laser engraving. There are several methods for achieving this, each with its own advantages and limitations.

- **Manual Tracing:** This involves manually tracing the outlines of a raster image using a vector graphics editor. While it offers precise control over the resulting vector image, it can be time-consuming for complex images.
- **Image Tracing Software:** Many design software programs offer built-in image tracing tools that can automatically convert raster images to vector format. These tools analyze the image and create vector paths based on the outlines and shapes detected.

- **Specialized Conversion Software:** There are specialized software programs designed specifically for converting raster images to vector format. These programs often offer advanced features and algorithms that can produce high-quality vector images from even complex raster images.

The best method for converting images for laser engraving will depend on the complexity of the image, the desired level of detail, and your personal preferences. Experiment with different methods to find the one that works best for you.

4.5 Importing and Exporting Files

Once you've created your design, you'll need to import it into your laser engraving software to send it to the machine. Most laser engraving software supports a variety of file formats, including:

- **SVG (Scalable Vector Graphics):** A standard format for vector images that is widely supported by laser engraving software.
- **DXF (Drawing Exchange Format):** Another common format for vector images that is often used for exchanging designs between different CAD and design software.
- **AI (Adobe Illustrator):** The native format for Adobe Illustrator files.
- **CDR (CorelDRAW):** The native format for CorelDRAW files.

- **PNG (Portable Network Graphics):** A raster image format that supports transparency and is often used for web graphics.
- **JPG (Joint Photographic Experts Group):** A common raster image format used for photographs and other images.

When importing files, it's important to ensure that the file format is compatible with your laser engraving software. If not, you may need to convert the file to a supported format before importing.

Similarly, when exporting files, choose a file format that is appropriate for the intended use. For example, if you're sharing a design with another person, you might choose a universal format like SVG or DXF. If you're printing the design, you might choose a high-resolution format like PNG or JPG.

4.6 Tips for Effective Designs

Creating effective designs for laser engraving involves more than just artistic talent. It also requires technical knowledge and an understanding of the limitations and possibilities of the laser engraving process. Here are some tips to help you create designs that translate well to laser engravings:

- **Line Thickness:** Ensure that the lines in your design are thick enough to be engraved clearly. Thin lines may not engrave properly or may be difficult to see on the finished product.

- **Detail Level:** Avoid overly intricate details that might get lost in the engraving process. Focus on clean lines and simple shapes that will translate well to the material.
- **Contrast:** Create contrast between the engraved areas and the background to make your design pop. Use contrasting colors, textures, or depths to enhance visibility.
- **Material Considerations:** Choose materials that are suitable for laser engraving and that complement your design. Different materials have different engraving characteristics, so experiment to find the best combinations.
- **Safety:** Be mindful of safety considerations when designing for laser engraving. Avoid using materials that release harmful fumes or are prone to catching fire. Always wear safety glasses and work in a well-ventilated area.
- **Test Runs:** Before engraving your final design, always perform test runs on scrap material to verify the settings and ensure the desired results. This will help you avoid costly mistakes and ensure the best possible outcome.

4.7 Common Design Mistakes to Avoid

Even experienced designers can make mistakes when creating designs for laser engraving. Here are some common pitfalls to avoid:

- **Overly Complex Designs:** As mentioned earlier, overly complex designs with intricate details may not translate well to laser

engraving. Keep your designs simple and focus on clean lines and shapes.
- **Thin Lines:** Thin lines may not engrave properly or may be difficult to see on the finished product. Ensure that the lines in your design are thick enough to be clearly visible.
- **Insufficient Contrast:** If there isn't enough contrast between the engraved areas and the background, your design may not stand out. Use contrasting colors, textures, or depths to create visual interest.
- **Ignoring Material Properties:** Different materials have different engraving characteristics. Ignoring these properties can lead to disappointing results. Always test your design on scrap material before engraving your final project.
- **Not Proofreading:** Typos and spelling errors can ruin an otherwise beautiful engraving.

CHAPTER 5

LASER ENGRAVING BASICS

Welcome to the heart of laser engraving action! In this chapter, we'll delve into the essential techniques and settings that will bring your designs to life on various materials. We'll unravel the mysteries of laser settings, guide you through choosing the right parameters for different materials, and explore the nuances of engraving text, images, and intricate patterns. By the end of this chapter, you'll have a solid grasp of the fundamental skills needed to operate your laser engraver with confidence and precision. So, let's ignite the laser beam and embark on this exciting journey of discovery!

5.1 Understanding Laser Settings (Power, Speed, DPI)

The key to successful laser engraving lies in understanding the interplay between three crucial settings: power, speed, and DPI (dots per inch). These settings determine the laser beam's intensity, movement speed, and the resolution of the engraving, respectively. Mastering these settings is essential for achieving the desired results on your chosen materials.

- **Power:** The power setting controls the laser beam's intensity, which directly impacts the depth and darkness of the engraving. Higher power settings result in deeper and darker engravings, while lower power settings produce shallower and lighter markings. The appropriate power setting depends on the material you're engraving and the desired outcome. For example, delicate materials like paper or thin fabrics require lower power settings to avoid burning, while denser materials like wood or acrylic can handle higher power settings.
- **Speed:** The speed setting determines how quickly the laser beam moves across the material. Faster speeds result in shallower engravings, while slower speeds produce deeper markings. The ideal speed setting depends on the material's thickness, density, and the desired level of detail in the engraving. For example, intricate designs may require slower speeds to ensure accuracy, while larger areas can be engraved faster without sacrificing quality.
- **DPI (Dots Per Inch):** DPI refers to the resolution of the engraving, which determines the level of detail and clarity. Higher DPI values produce finer details and smoother gradients, while lower DPI values result in coarser engravings with less detail. The choice of DPI depends on the complexity of your design and the intended purpose of the engraving. For example, high DPI values are

essential for photorealistic engravings, while lower DPI values may be sufficient for simple text or logos.

Finding the optimal combination of power, speed, and DPI for your specific material and design requires experimentation and testing. Start with conservative settings and gradually adjust them until you achieve the desired results. It's also helpful to refer to the manufacturer's recommended settings for your specific laser engraver and material.

5.2 Choosing the Right Settings for Different Materials

Each material responds differently to laser engraving, requiring specific settings to achieve the best results. Here are some general guidelines for choosing the right settings for common materials:

- **Wood:** Wood is a versatile material that can be engraved with a wide range of power and speed settings. Start with a medium power setting and adjust it based on the desired depth of engraving. Slower speeds are recommended for intricate designs and finer details.
- **Metal:** Metal requires higher power settings than wood or acrylic due to its density and reflectivity. Experiment with different power and speed combinations to find the optimal settings for your specific metal type and thickness.

- **Acrylic:** Acrylic is a popular choice for laser engraving due to its versatility and ease of use. It can be engraved with moderate power and speed settings. However, it's important to avoid overheating the material, as this can cause it to melt or warp.
- **Leather:** Leather is a delicate material that requires lower power settings to avoid burning. Slower speeds are recommended to ensure precise and detailed engravings.
- **Glass:** Glass engraving requires a special technique known as "wet engraving," where a thin layer of water or other liquid is applied to the surface to prevent cracking. Use low power and high speed settings to avoid overheating the glass.
- **Fabric:** Certain fabrics, like cotton and denim, can be laser engraved. Use low power and high speed settings to avoid burning or scorching the fabric.

Remember, these are just general guidelines. The optimal settings for your specific project will depend on various factors, including the material type, thickness, color, and the desired engraving effect. Always test your settings on a scrap piece of material before engraving your final project.

5.3 Engraving Text

Engraving text with a laser engraver is a common and versatile application. Whether you're personalizing gifts, creating signage, or adding artistic flair to objects, engraved text can add a touch of sophistication and elegance. Here are some tips for engraving text effectively:

- **Choose the Right Font:** Select a font that is legible and appropriate for the purpose of the engraving. Avoid using overly decorative fonts that might be difficult to read. Consider the size and style of the font in relation to the size of the object being engraved.
- **Adjust Line Spacing:** Adjust the spacing between lines of text to ensure readability and visual appeal. Too much spacing can make the text appear disjointed, while too little spacing can make it difficult to read.
- **Utilize Kerning:** Kerning is the process of adjusting the spacing between individual letters to improve the overall appearance and readability of the text. Some design software programs offer automatic kerning tools, while others require manual adjustment.
- **Consider Orientation:** Decide on the orientation of the text—horizontal, vertical, or angled—based on the shape and size of the object being engraved. Ensure that the text is positioned in a way that is both aesthetically pleasing and easy to read.
- **Test and Adjust:** Before engraving your final text, always perform test runs on scrap material to verify the settings and ensure the desired results. Adjust the power, speed, and DPI settings as needed to achieve the desired depth and clarity.

5.4 Engraving Images

Laser engraving images opens up a world of creative possibilities. You can engrave photographs, artwork, logos, or any other visual design onto a variety of materials. Here are some tips for engraving images effectively:

- **Choose the Right Image:** Select an image with good contrast and sharp details. High-resolution images will generally produce better results than low-resolution images. Avoid images with complex gradients or subtle color variations, as these can be difficult to reproduce accurately with laser engraving.
- **Convert to Grayscale:** Most laser engravers work in grayscale, so it's important to convert your image to grayscale before engraving. This will ensure that the different shades of gray in the image are accurately represented in the engraving.
- **Adjust Image Settings:** Depending on your design software and laser engraver, you may have options to adjust the image's brightness, contrast, and sharpness. Experiment with these settings to optimize the image for laser engraving.
- **Dithering:** If your image has a lot of detail or subtle gradations, you may want to consider using dithering. Dithering is a technique that simulates shades of gray by using patterns of black and white dots. This can help to create more realistic and detailed engravings.

- **Test and Adjust:** As with text engraving, always perform test runs on scrap material before engraving your final image. Adjust the power, speed, and DPI settings as needed to achieve the desired results.

5.5 Creating Depth and Dimension

Laser engraving can be used to create not only flat markings but also engravings with depth and dimension. This can add a tactile element to your designs and make them more visually interesting. Here are some techniques for creating depth and dimension:

- **Varying Power Settings:** By adjusting the power setting during the engraving process, you can create engravings with varying depths. For example, you could use a higher power setting for the foreground elements of a design and a lower power setting for the background elements, creating a sense of depth and perspective.
- **Multiple Passes:** Multiple passes with the laser beam can also be used to create deeper engravings. By repeating the engraving process multiple times, you can gradually remove more material and achieve a greater depth.
- **3D Engraving:** Some laser engravers have the capability to create 3D engravings. This involves using specialized software to generate a 3D model of your design, which is then

translated into a series of laser engraving paths. The laser engraver follows these paths to create a three-dimensional engraving on the material.

5.6 Testing and Adjusting Settings

Testing and adjusting settings is a crucial part of the laser engraving process. Before engraving your final project, always perform test runs on scrap material to verify the settings and ensure the desired results. This allows you to fine-tune the power, speed, and DPI settings without risking damage to your final material.

When testing, start with conservative settings and gradually increase the power and speed until you achieve the desired depth and clarity. Pay attention to how the material responds to the laser beam and adjust the settings accordingly. If you notice any burning, charring, or melting, reduce the power or increase the speed.

Keep a record of the settings you use for each material and design. This will save you time in the future and help you achieve consistent results.

5.7 Troubleshooting Common Engraving Issues

Even with careful planning and testing, you may encounter some common issues during the laser engraving process. Here are some troubleshooting tips for common problems:

- **Uneven Engraving:** This can be caused by an uneven surface, inconsistent material thickness, or improper focus of the laser beam. Check the material for any inconsistencies and ensure that it is securely fastened to the engraver's bed. Adjust the focus of the laser beam if necessary.
- **Burning or Charring:** This can occur when the laser power is too high or the speed is too slow for the material. Reduce the power or increase the speed to avoid burning. You can also try using masking tape to protect the material from the laser beam.

Fuzzy or Blurry Engravings: This can be caused by a dirty lens or mirror, improper focus, or vibrations during the engraving process. Clean the lens or mirror with a soft cloth and specialized cleaning solution. Check for loose parts or vibrations in the engraver and tighten or adjust them as needed.

- **Inconsistent Engraving Depth:** This can be caused by variations in the material thickness or inconsistent power delivery from the laser. Check the material for thickness variations and

ensure that it is level on the bed. If the power delivery is inconsistent, there may be a problem with the laser engraver itself, and you may need to consult the manufacturer's support.
- **Skipped Lines or Areas:** This can occur if the speed setting is too high for the laser to accurately follow the design. Reduce the speed setting to allow the laser more time to engrave each line or area.
- **Material Damage:** This can happen when the laser power is too high for the material or when the material is not properly prepared. Always test your settings on a scrap piece of material before engraving your final project. Use masking tape or other protective measures to shield delicate materials from the laser beam.

By understanding these common issues and their potential causes, you can quickly diagnose and resolve problems that may arise during the laser engraving process. Don't be discouraged if you encounter setbacks; troubleshooting is a natural part of the learning process. With patience and persistence, you'll soon be able to overcome challenges and achieve consistently excellent results.

Mastering the basics of laser engraving is a rewarding journey that opens up a world of creative possibilities. By understanding the interplay of laser settings, choosing the right parameters for different materials, and honing your engraving techniques, you can create stunning and personalized works of art.

Remember, practice makes perfect. The more you experiment with different materials, settings, and techniques, the more proficient you'll become. Don't be afraid to push the boundaries of your creativity and explore the endless possibilities that laser engraving offers.

CHAPTER 6

ADVANCED LASER ENGRAVING TECHNIQUES

Welcome to the realm of laser engraving mastery! In this chapter, we'll venture beyond the basics and explore advanced techniques that will elevate your engraving skills to new heights. We'll delve into the intricacies of engraving on curved surfaces, crafting mesmerizing 3D engravings, seamlessly combining engraving and cutting, mastering the art of photo engraving, infusing multiple colors into your designs, experimenting with captivating fill patterns and textures, and ultimately, encouraging you to embrace experimentation and discover your unique artistic style. So, fasten your seatbelts and prepare to unlock a world of limitless possibilities with advanced laser engraving!

6.1 Engraving on Curved Surfaces

Engraving on flat surfaces is a great starting point, but the real magic happens when you conquer the challenge of engraving on curved surfaces. This technique adds a new dimension to your creations, allowing you to personalize cylindrical objects, spherical items, or any other non-flat surface.

To achieve successful engravings on curved surfaces, there are several approaches you can employ:

- **Rotary Attachment:** This specialized accessory for laser engravers allows you to rotate cylindrical objects like glasses, mugs, or bottles during the engraving process. The rotary attachment ensures that the laser beam remains focused on the curved surface, resulting in a seamless and even engraving.
- **4th Axis Rotary:** A 4th axis rotary is an advanced rotary attachment that enables engraving on both cylindrical and spherical objects. It offers more flexibility and precision, making it ideal for intricate designs and complex shapes.
- **Software Compensation:** Some laser engraving software programs offer features that compensate for the curvature of the surface. This involves adjusting the laser path to account for the changing distance between the laser head and the material's surface.
- **Manual Adjustments:** In some cases, you may need to manually adjust the focus of the laser beam as you engrave on a curved surface. This requires practice and skill, but it can be a valuable technique for achieving precise results on unique or irregular shapes.

Whichever method you choose, it's important to start with simple designs and test your settings on scrap material before engraving your final project. Engraving on curved surfaces requires patience and attention to detail, but the results can be truly stunning.

6.2 Creating 3D Engravings

3D laser engraving is a fascinating technique that adds depth and realism to your designs. Unlike traditional 2D engraving, which only creates surface markings, 3D engraving creates a relief pattern that you can feel and see. This technique is perfect for creating intricate sculptures, personalized gifts, and unique decorative items.

There are two main approaches to creating 3D engravings:

- **Greyscale Engraving:** This technique involves using shades of gray to represent different depths in the design. The laser engraver adjusts the power of the beam based on the shade of gray, resulting in varying depths of engraving. This technique is ideal for creating subtle reliefs and organic shapes.
- **3D Modeling Software:** For more complex and detailed 3D engravings, specialized 3D modeling software is used to create a digital model of the design. This model is then sliced into layers, and the laser engraver follows these layers to create the 3D relief.

3D engraving requires specialized software and techniques, but the results can be truly breathtaking.

With practice and experimentation, you can create stunning 3D engravings that will impress and amaze.

6.3 Combining Engraving and Cutting

Laser engraving and laser cutting are two complementary processes that can be combined to create intricate and unique designs. By selectively engraving and cutting different parts of a material, you can create multi-layered designs, intricate patterns, and even functional objects.

For example, you could engrave a design onto a piece of wood and then cut out the shape of the design, creating a decorative plaque or sign. You could also engrave a pattern onto a sheet of acrylic and then cut it into various shapes to create custom keychains or ornaments.

Combining engraving and cutting requires careful planning and design. You'll need to consider the order of operations, the settings for each process, and how the two processes will interact with each other. With practice, you can master this technique and create truly unique and personalized projects.

6.4 Photo Engraving

Photo engraving is the process of using a laser engraver to recreate a photograph or other image onto a material. This technique can be used to create personalized gifts, memorials, awards, and other decorative items.

Photo engraving requires special software and techniques to achieve optimal results. The image is typically converted to grayscale and then dithered to create a pattern of dots that simulates the shades of gray in the original image. The laser engraver then follows this pattern to etch the image onto the material.

Achieving high-quality photo engravings can be challenging, but with practice and the right tools, you can create stunning results. Experiment with different materials, settings, and dithering patterns to find the optimal combination for your specific project.

6.5 Engraving with Multiple Colors

Laser engraving is not limited to grayscale or monochrome designs. With the right techniques and materials, you can create engravings with multiple colors. This can add a vibrant and eye-catching element to your projects.

There are a few different ways to achieve multi-color laser engraving:

- **Using Colored Materials:** One approach is to use materials that are already colored, such as anodized aluminum or two-tone acrylic. The laser engraver removes the top layer of the material, revealing the different colors underneath.
- **Applying Color Fill:** Another method is to engrave a design onto a material and then fill

the engraved areas with colored paint or resin. This technique can be used on various materials, including wood, metal, and acrylic.
- **MOPA Fiber Lasers:** MOPA (Master Oscillator Power Amplifier) fiber lasers are a newer technology that allows for color marking on certain metals. This technique involves adjusting the laser's pulse duration and frequency to create different colors on the metal surface.

Engraving with multiple colors requires careful planning and execution, but the results can be truly spectacular. With creativity and experimentation, you can create stunning designs that pop with color.

6.6 Using Fill Patterns and Textures

Fill patterns and textures can add depth, dimension, and visual interest to your laser engravings. Instead of simply filling an engraved area with a solid color, you can use a pattern or texture to create a more unique and eye-catching effect.

There are countless fill patterns and textures available, ranging from simple geometric shapes to complex organic patterns. You can create your own custom patterns or use pre-made patterns available in design software or online libraries.

When choosing a fill pattern or texture, consider the overall style of your design and the material you're using. Some patterns may work better on certain materials than others. For example, a fine, detailed pattern may look great on a smooth surface like metal, while a bolder pattern may be more suitable for a textured surface like wood.

Experiment with different fill patterns and textures to discover new possibilities and add a unique touch to your engravings.

6.7 Experimenting with Different Techniques

One of the most exciting aspects of laser engraving is the endless potential for experimentation. Don't be afraid to try new techniques, materials, and designs. Push the boundaries of your creativity and see what you can create.

Here are some ideas for experimentation:

- **Combining different materials:** Try engraving on multiple materials layered together to create unique and interesting effects.
- **Incorporating other techniques:** Combine laser engraving with other techniques like painting, staining, or epoxy resin to create truly unique pieces.
- **Playing with different settings:** Experiment with different power, speed, and DPI settings to achieve different engraving depths, textures, and finishes.
- **Creating your own designs:** Don't be limited by pre-made designs. Use your imagination

and design software to create your own unique patterns, logos, and artwork.

Remember, the key to mastering laser engraving is practice and experimentation. The more you try new things, the more you'll learn and grow as an artist.

Advanced laser engraving techniques open up a world of creative possibilities. By mastering these techniques, you can create stunning and unique projects that go beyond the ordinary. From engraving on curved surfaces to creating 3D engravings, combining engraving and cutting, and experimenting with different materials and techniques, the only limit is your imagination.

Embrace the challenge of learning new skills and experimenting with different approaches. With dedication and practice, you can achieve incredible results and take your laser engraving to the next level. So, go forth and unleash your creativity!

CHAPTER 7

LASER CUTTING

Welcome to the realm of laser cutting, where precision meets power! In this chapter, we'll shift our focus from engraving to the equally captivating art of laser cutting. We'll unravel the secrets behind laser cutting settings, guide you through selecting the ideal parameters for various materials, and explore techniques for cutting both simple and intricate shapes. We'll delve into the intricacies of cutting through thick materials and share invaluable tips for achieving unparalleled precision. By the end of this chapter, you'll wield the laser beam like a digital scalpel, effortlessly transforming materials into bespoke creations. So, let's dive in and discover the transformative power of laser cutting!

7.1 Understanding Laser Cutting Settings

Laser cutting, like engraving, relies on a delicate balance of settings to achieve optimal results. These settings determine how the laser beam interacts with the material, ultimately dictating the quality and precision of the cut. Let's break down the key settings you need to understand:

- **Power:** The power setting controls the intensity of the laser beam, influencing the speed and

depth of the cut. Higher power settings allow for faster cutting and can penetrate thicker materials, while lower power settings are ideal for delicate materials or intricate cuts. The optimal power setting depends on the material's thickness, density, and desired cutting speed.
- **Speed:** The speed setting determines how quickly the laser beam traverses the material. Faster speeds are suitable for simpler cuts and thinner materials, while slower speeds are necessary for thicker materials or intricate designs. Striking the right balance between power and speed is crucial for achieving clean and precise cuts without compromising on efficiency.
- **Frequency:** The frequency setting dictates how rapidly the laser beam pulses on and off. Higher frequencies generally result in smoother cuts with less charring, while lower frequencies might be suitable for specific materials or applications. Experimentation with frequency settings can help you fine-tune the cut quality for different materials and desired outcomes.
- **Air Assist:** Air assist is a crucial component in laser cutting, as it directs a stream of air onto the cutting area. This airflow helps to remove debris, cool the material, and prevent flare-ups, ensuring clean and consistent cuts. The pressure and direction of the air assist can be adjusted based on the material and cutting requirements.

- **Focus:** The focus of the laser beam plays a vital role in cutting accuracy. A properly focused beam concentrates the laser's energy into a small spot, resulting in cleaner and more precise cuts. The focal point should be adjusted based on the material's thickness and the desired cutting depth.

Mastering these laser cutting settings requires a combination of technical knowledge and practical experience. Always refer to your laser cutter's manual for recommended settings and don't hesitate to experiment with different parameters on scrap material before tackling your final project.

7.2 Choosing the Right Settings for Different Materials

Each material possesses unique characteristics that influence its interaction with the laser beam during cutting. Understanding these characteristics is paramount for selecting the appropriate settings and achieving the desired outcomes. Let's explore some common materials and their corresponding laser cutting settings:

- **Wood:** Wood is a popular choice for laser cutting due to its versatility and natural appeal. When cutting wood, a medium power setting and moderate speed are often a good starting point. However, adjusting the settings based on the wood species, thickness, and desired level of detail is crucial. Hardwood

typically requires higher power settings than softwood, while thicker pieces necessitate slower cutting speeds.
- **Acrylic:** Acrylic is another versatile material widely used in laser cutting. It responds well to moderate power and speed settings, producing clean and polished edges. However, excessive heat can cause acrylic to melt or warp, so it's important to maintain a consistent cutting speed and use air assist to dissipate heat.
- **Leather:** Leather is a delicate material that requires careful attention during laser cutting. A low to medium power setting and slow cutting speed are generally recommended to prevent burning or charring. Air assist is essential to minimize smoke and residue buildup.
- **Paper and Cardboard:** These materials are relatively thin and easy to cut with laser cutters. Lower power and faster speed settings are typically sufficient. However, it's important to monitor the cutting process closely to avoid excessive burning or scorching.
- **Fabric:** Certain fabrics, like cotton and felt, can be laser cut to create intricate patterns and designs. Lower power settings and moderate to high speeds are generally used, depending on the fabric's thickness and density. Air assist is essential to prevent the fabric from fraying or catching fire.
- **Metal:** While most desktop laser cutters are not equipped to cut metal, industrial laser cutters

can effortlessly slice through various metals. The power and speed settings for metal cutting vary significantly depending on the type of metal and its thickness. Specialized laser cutters with higher power outputs and oxygen assist are often used for metal cutting applications.

Remember, these are just general guidelines. The optimal settings for your specific project will depend on the material type, thickness, color, and desired cutting quality. Always consult the material manufacturer's recommendations and conduct test cuts on scrap material before finalizing your settings.

7.3 Cutting Basic Shapes

Laser cutting excels at producing precise and consistent basic shapes like squares, rectangles, circles, and triangles. Whether you're creating stencils, custom labels, or geometric patterns, cutting basic shapes is a fundamental skill in laser cutting.

To cut basic shapes, you'll typically start by designing your shapes in a vector graphics editor like Inkscape or Adobe Illustrator. Once your design is ready, import it into your laser cutter software. Most laser cutter software allows you to set cutting parameters for each shape individually, giving you precise control over the cutting process.

Before cutting, ensure that your material is securely placed on the laser cutter's bed and that the focus of the laser beam is properly adjusted. Start with conservative power and speed settings and gradually increase them until you achieve the desired cut quality.

For simple shapes, a single pass with the laser beam is usually sufficient. However, for thicker materials or more intricate shapes, multiple passes with lower power settings might be necessary to achieve clean and precise cuts without burning or warping the material.

7.4 Cutting Complex Designs

Laser cutting truly shines when it comes to creating complex and intricate designs. With its precision and versatility, laser cutting can bring even the most elaborate patterns and shapes to life.

To cut complex designs, you'll typically follow a similar workflow as cutting basic shapes. However, there are a few additional considerations to keep in mind:

- **Node Editing:** If your design has intricate details or tight corners, you may need to use node editing tools in your design software to refine the vector paths. This will ensure that the laser cutter can accurately follow the design and produce clean cuts.
- **Lead-in and Lead-out Points:** Adding lead-in and lead-out points to your design can help prevent scorching or burning at the start and

end of each cut. These points are small extensions of the cut path that allow the laser beam to ramp up and down in power gradually.
- **Bridge Cutting:** For delicate or intricate designs, consider using bridge cutting. This technique involves leaving small bridges of material between cut sections to provide support and prevent the design from falling apart during cutting. The bridges can be easily removed after cutting is complete.
- **Multiple Passes:** For thicker materials or deeper cuts, multiple passes with the laser beam might be necessary. Start with lower power settings and gradually increase them with each pass to achieve the desired depth without compromising cut quality.
- **Layering:** For complex designs with multiple layers or overlapping elements, consider cutting each layer separately and then assembling them. This can simplify the cutting process and ensure accurate alignment of different design elements.

Cutting complex designs requires patience, precision, and a good understanding of your laser cutter's capabilities. With practice and experimentation, you can create stunning and intricate pieces that showcase the true potential of laser cutting.

7.5 Cutting through Thick Materials

While most desktop laser cutters are designed for cutting thinner materials like paper, acrylic, or thin wood, some models can handle thicker materials with the right settings and techniques.

To cut through thick materials, you'll generally need to use higher power settings and slower cutting speeds. However, increasing the power too much can cause burning or charring, while excessively slow speeds can lead to overheating and warping. It's crucial to strike the right balance between power and speed for each material and thickness.

Here are some tips for cutting through thick materials:

- **Multiple Passes:** For very thick materials, multiple passes with lower power settings might be necessary. Each pass will gradually cut deeper into the material until the desired thickness is reached.
- **Focus Adjustment:** The focus of the laser beam should be adjusted based on the material's thickness. Thicker materials typically require a slightly higher focal point to ensure optimal cutting performance.
- **Air Assist:** A strong air assist is essential for cutting through thick materials. It helps to remove debris, cool the material, and prevent flare-ups, ensuring clean and consistent cuts.

- **Material Preparation:** Properly preparing the material can make a significant difference in cutting quality. Ensure that the material is clean, flat, and securely fastened to the laser cutter's bed. If the material has any warping or unevenness, you may need to sand or flatten it before cutting.

Cutting through thick materials can be challenging, but with the right techniques and settings, you can achieve impressive results. Remember to experiment and test different settings on scrap material before cutting your final project.

7.6 Tips for Precise Cutting

Achieving precise and accurate cuts is the hallmark of laser cutting. Here are some tips to help you achieve the highest level of precision:

- **Clean Optics:** Ensure that the laser cutter's mirrors, lenses, and other optical components are clean and free of dust or debris. Dirty optics can scatter the laser beam and affect cutting accuracy.

- **Secure Material:** Ensure that your material is firmly held in place on the laser cutter's bed to prevent any movement during cutting. Use clamps, tape, or other suitable methods to secure the material without obstructing the laser path.

- **Calibrate Laser Path:** Some laser cutters require calibration of the laser path to ensure accurate positioning and movement of the laser head. Refer to your laser cutter's manual for instructions on how to perform this calibration.
- **Optimize Cutting Path:** In your design software, optimize the cutting path to minimize unnecessary movements and reduce cutting time. For example, group shapes that can be cut together and arrange the cutting sequence in a way that minimizes travel distance between cuts.
- **Use Sharp Blades:** If your laser cutter uses a blade to pierce the material before cutting, ensure that the blade is sharp and clean. A dull blade can result in ragged or uneven cuts.
- **Monitor Cutting Process:** Keep a close eye on the cutting process and be prepared to pause or stop the machine if you notice any issues, such as excessive burning or deviations from the design.
- **Test Cuts:** Always perform test cuts on scrap material before committing to your final project. This allows you to fine-tune settings, identify potential problems, and ensure the desired results.

By following these tips and paying attention to detail, you can achieve remarkably precise cuts that elevate the quality and aesthetics of your laser cutting projects.

7.7 Troubleshooting Common Cutting Issues

Even with meticulous planning and execution, you might encounter some common challenges during laser cutting. Here are some troubleshooting tips for addressing typical issues:

- **Inconsistent Cut Depth:** If the cut depth varies throughout your project, it could be due to uneven material thickness, improper focus, or inconsistent power delivery from the laser. Check your material for any variations in thickness and ensure it's level on the bed. Adjust the focus if needed, and if the issue persists, consult the laser cutter's manual or seek assistance from the manufacturer.
- **Jagged or Ragged Edges:** Jagged edges can result from a dull blade, incorrect cutting speed, or vibrations in the machine. Ensure your blade is sharp, adjust the cutting speed to a slower pace, and check for any loose components or vibrations in the laser cutter.
- **Burning or Charring:** Excessive burning or charring can occur when the laser power is too high or the cutting speed is too slow. Reduce the power setting and increase the speed to minimize burning.
- **Material Warping:** Some materials, like acrylic, are prone to warping due to heat generated during laser cutting. To prevent warping, use air assist to dissipate heat, adjust the cutting

speed to avoid overheating, and consider using a backing material to support the workpiece.
- **Incomplete Cuts:** If the laser fails to cut through the material completely, it could be due to insufficient power, incorrect focus, or a dirty lens or mirror. Increase the power setting, adjust the focus, and clean the optical components as needed.
- **Flare-ups:** Flare-ups occur when the material being cut ignites or produces excessive smoke. To prevent flare-ups, use air assist to blow away debris and fumes, adjust the cutting speed to avoid overheating, and ensure the material is properly secured.
- **Software Issues:** If you encounter problems with the laser cutter software, such as incorrect positioning or unexpected movements, try restarting the software and the laser cutter. If the issue persists, consult the software's documentation or contact the manufacturer for support.

Troubleshooting laser cutting issues requires a systematic approach and a willingness to experiment. By identifying the root cause of the problem and applying the appropriate solution, you can overcome challenges and achieve consistently excellent cutting results.

Laser cutting is a transformative technology that empowers you to turn raw materials into exquisite creations. By understanding the intricacies of laser cutting settings, choosing the right parameters for different materials, and mastering techniques for cutting various shapes, you can unlock a world of creative possibilities.

Embrace the versatility and precision of laser cutting to craft personalized gifts, intricate artwork, functional prototypes, and so much more. With practice and a thirst for innovation, you'll soon be wielding the laser beam like a true artisan, transforming your ideas into tangible masterpieces. So, let your imagination soar and embark on a journey of laser cutting excellence!

CHAPTER 8

CREATING PROJECTS

Welcome to the realm of boundless creativity! In this chapter, we embark on an exciting journey into the world of project creation with your laser engraver. By now, you've mastered the technical aspects of laser engraving and cutting, and you're ready to unleash your imagination. We'll guide you through the entire project lifecycle, from brainstorming innovative ideas to designing, planning, executing, and finally showcasing your unique creations. Whether you're crafting personalized gifts, designing intricate artwork, or building functional prototypes, this chapter will empower you to transform your ideas into tangible reality. So, let's dive into the exhilarating process of bringing your projects to life!

8.1 Brainstorming Project Ideas

The first step in any creative endeavor is to brainstorm ideas. This is where you let your imagination run wild and explore the endless possibilities that laser engraving and cutting offer. Here are some tips to spark your creativity:

- **Explore Your Interests:** Start by thinking about your passions, hobbies, and interests. What kind of projects would you enjoy creating? Do you have a favorite artist or style that inspires you? Brainstorming ideas that resonate with

you personally will make the creative process more enjoyable and fulfilling.
- **Consider Your Skills:** While laser engraving and cutting are relatively easy to learn, certain projects may require more advanced techniques or materials. Be realistic about your current skill level and choose projects that are challenging but achievable.
- **Research Online:** The internet is a treasure trove of laser engraving and cutting project ideas. Explore online platforms like Pinterest, Instagram, and YouTube for inspiration. Look for tutorials, project showcases, and community forums where you can connect with other laser enthusiasts and exchange ideas.
- **Think Outside the Box:** Don't limit yourself to conventional project ideas. Think outside the box and explore unique and unconventional applications of laser engraving and cutting. Consider combining different materials, techniques, and designs to create truly original pieces.
- **Solve Problems:** Identify problems or challenges in your everyday life that could be solved with laser engraving or cutting. This could be anything from creating custom organizers for your workspace to designing personalized labels for your belongings.
- **Collaborate with Others:** Brainstorming with friends, family, or fellow laser enthusiasts can spark new ideas and perspectives. Bounce ideas off each other, share inspiration, and

collaborate on projects to achieve even greater results.

Remember, the best ideas often come from unexpected places. Be open to inspiration from all sources, and don't be afraid to experiment and try new things. The possibilities are truly limitless!

8.2 Designing and Planning Your Projects

Once you have a few project ideas in mind, it's time to start designing and planning. This stage involves translating your creative vision into a tangible blueprint for your laser engraver.

- **Sketching:** Start by sketching your ideas on paper or using a digital drawing tablet. Don't worry about perfection at this stage; focus on capturing the essence of your design and exploring different layouts and compositions.
- **Choosing Materials:** Select the materials that best suit your project. Consider factors like the desired aesthetic, functionality, durability, and compatibility with laser engraving and cutting. Refer to Chapter 3 for a comprehensive guide on choosing and preparing materials.
- **Creating Digital Designs:** Use your chosen design software to create digital versions of your sketches. Incorporate the design principles discussed in Chapter 4 to ensure your designs are visually appealing and effective. Experiment with different fonts,

colors, patterns, and textures to achieve the desired look and feel.
- **Determining Laser Settings:** Based on your chosen materials and design, determine the appropriate laser settings for engraving and cutting. Refer to the guidelines in Chapters 5 and 7 for recommended settings for different materials.
- **Creating a Project Plan:** Break down your project into smaller, manageable tasks. This will help you stay organized and ensure that you have all the necessary materials and tools on hand before you start engraving or cutting.
- **Safety Considerations:** Always prioritize safety when designing and planning your projects. Refer to the safety guidelines in Chapter 1 and any specific precautions for the materials you'll be using.

By carefully designing and planning your projects, you set yourself up for success and ensure that your creative vision is executed smoothly and efficiently.

8.3 Gathering Materials and Tools

Before you start engraving or cutting, it's important to gather all the necessary materials and tools. This will prevent delays and ensure that you have everything you need to complete your project without interruptions.

- **Materials:** Gather all the materials you'll be using for your project, including the main

material for engraving or cutting, any additional materials for layering or embellishments, and any adhesives or fasteners you might need.
- **Tools:** In addition to your laser engraver, you may need other tools like a ruler, measuring tape, cutting mat, masking tape, clamps, pliers, screwdrivers, and safety glasses. Refer to your project plan to identify any specific tools required for your project.
- **Personal Protective Equipment (PPE):** Always prioritize safety by wearing appropriate PPE, including safety glasses and gloves. Depending on the materials you're using, you may also need a respirator or other protective gear.
- **Cleaning Supplies:** Have cleaning supplies on hand to clean your materials before and after engraving or cutting. This will help prevent debris from interfering with the laser beam and ensure a clean finish for your project.

By having all your materials and tools organized and readily available, you can streamline your workflow and focus on the creative process.

8.4 Executing Your Projects Step-by-Step

Now comes the exciting part – executing your projects! This is where your designs come to life through the power of laser engraving and cutting. Here's a general step-by-step guide to executing your projects:

1. **Prepare Your Materials:** Clean and prepare your materials according to the guidelines in Chapter 3. Apply masking tape or other protective measures if necessary.
2. **Secure Materials:** Secure your materials to the laser engraver's bed using clamps or tape to prevent any movement during engraving or cutting.
3. **Import Design:** Import your digital design into the laser engraver software.
4. **Adjust Settings:** Adjust the laser settings based on your chosen materials and design. Refer to the guidelines in Chapters 5 and 7 for recommended settings.
5. **Test Run:** Always perform a test run on a scrap piece of material before engraving or cutting your final project. This will help you verify the settings and identify any potential issues.
6. **Engrave or Cut:** Once you're satisfied with the test run, proceed to engrave or cut your final project. Monitor the process closely and adjust the settings or positioning if necessary.
7. **Clean Up:** After engraving or cutting, clean up any debris or residue from your materials and tools.
8. **Finishing Touches:** Apply any finishing touches, such as painting, staining, or sealing, according to your project plan.

Remember to take your time, be patient, and enjoy the process of creating something unique and personal.

Don't be afraid to experiment and try new techniques as you go along.

8.5 Finishing and Polishing Your Projects

The final step in creating a laser engraving or cutting project is finishing and polishing. This involves refining the edges, removing any imperfections, and applying any final touches that will enhance the overall appearance and quality of your creation.

- **Removing Masking:** Carefully remove any masking tape or other protective materials used during the engraving or cutting process.
- **Sanding:** Sand any rough edges or surfaces to create a smooth and polished finish. Use sandpaper with progressively finer grits to achieve the desired level of smoothness.
- **Cleaning:** Clean the project thoroughly to remove any dust, debris, or residue from the engraving or cutting process.
- **Applying Finishes:** Apply any finishes, such as paint, stain, sealant, or wax, according to your project plan. These finishes can enhance the appearance, durability, and longevity of your project.
- **Assembly:** If your project involves multiple parts, assemble them according to your plan. Use adhesives, fasteners, or other joining methods as needed.
- **Quality Control:** Inspect your finished project for any imperfections or defects. If necessary,

make any final adjustments or repairs before showcasing or gifting your creation.

By taking the time to finish and polish your projects, you elevate them from simple creations to polished masterpieces. A well-finished project not only looks more professional but also reflects the care and attention to detail you put into its creation.

8.6 Showcasing Your Creations

Once your projects are complete, it's time to share them with the world! Showcasing your laser engravings and cuttings can be a rewarding experience, allowing you to connect with others who share your passion for creativity and craftsmanship.

- **Online Platforms:** Share photos and videos of your projects on social media platforms like Instagram, Pinterest, and Facebook. Join online communities and forums dedicated to laser engraving and cutting to connect with fellow enthusiasts and share tips and inspiration.
- **Craft Fairs and Markets:** Participate in craft fairs and markets to showcase and sell your creations. This is a great way to get feedback from potential customers, network with other artisans, and build your brand.
- **Online Stores:** If you're looking to turn your hobby into a business, consider selling your laser engravings and cuttings online. Platforms like Etsy, Shopify, and Amazon Handmade

- offer easy-to-use tools for creating an online store and reaching a global audience.
- **Gifting:** Personalized laser engravings and cuttings make thoughtful and unique gifts for friends and family. Consider creating custom gifts for birthdays, holidays, or special occasions.
- **Personal Use:** Of course, you can also simply enjoy your creations for personal use. Display them in your home or office, use them as functional objects, or simply admire the beauty and craftsmanship you've achieved.

No matter how you choose to showcase your creations, the joy of bringing your ideas to life and sharing them with others is truly rewarding. Let your passion for laser engraving shine through, and inspire others with your unique and personalized projects.

8.7 Finding Inspiration Online

The internet is a vast and ever-evolving source of inspiration for laser engraving and cutting projects. With a few clicks, you can discover countless ideas, tutorials, and resources to fuel your creativity and expand your skills. Here are some online platforms and resources to explore:

- **Pinterest:** Pinterest is a visual discovery engine where you can find endless inspiration for laser engraving and cutting projects. Search for specific keywords like "laser engraving projects," "laser cut designs," or "DIY laser

crafts" to discover a wealth of ideas, from personalized gifts and home decor to intricate artwork and functional objects.
- **Instagram:** Instagram is a social media platform where you can connect with other laser engraving enthusiasts, artists, and businesses. Follow hashtags like #laserengraving, #lasercutting, and #makercommunity to discover inspiring projects and connect with like-minded creators.
- **YouTube:** YouTube is a vast repository of laser engraving and cutting tutorials, project showcases, and behind-the-scenes glimpses into the creative process. Search for channels dedicated to laser engraving and cutting to learn new techniques, discover project ideas, and get troubleshooting tips.
- **Online Forums and Communities:** Join online forums and communities dedicated to laser engraving and cutting to connect with fellow enthusiasts, ask questions, share your projects, and learn from others. These communities are valuable resources for troubleshooting problems, finding inspiration, and staying up-to-date on the latest trends and techniques.
- **Laser Engraving and Cutting Websites:** Many websites specialize in providing resources for laser engraving and cutting enthusiasts. These websites offer design templates, project ideas, tutorials, software recommendations, and even online marketplaces where you can sell your creations.

- **Manufacturer Websites:** Don't forget to explore the websites of laser engraver manufacturers. They often provide tutorials, project ideas, and support forums to help you get the most out of your machine.

By actively engaging with the online laser engraving and cutting community, you can expand your knowledge, discover new ideas, and connect with fellow creators who share your passion. The internet is a valuable resource for learning and inspiration, so don't hesitate to explore its vast offerings and fuel your creative fire.

Creating projects with your laser engraver is a thrilling journey that combines technical skills with artistic expression. From brainstorming innovative ideas to designing, planning, executing, and showcasing your creations, the entire process is a testament to your creativity and ingenuity.

Remember, the key to successful project creation is to embrace experimentation, explore new techniques, and push the boundaries of your imagination. Don't be afraid to try new things, learn from your mistakes, and most importantly, have fun along the way!

With the knowledge and resources shared in this chapter, you're well-equipped to embark on your own creative adventures. Whether you're crafting personalized gifts, designing intricate artwork, or building functional prototypes, the possibilities are endless. So, let your imagination soar and transform your ideas into tangible reality with the power of laser engraving and cutting!

CHAPTER 9

MAINTAINING YOUR LASER ENGRAVER

Welcome to the realm of laser engraver guardianship! In this chapter, we'll delve into the essential practices that will ensure your laser engraver remains a reliable and high-performing companion throughout your creative journey. We'll cover everything from regular cleaning and maintenance to replacing parts and consumables, troubleshooting common problems, and extending the lifespan of your beloved machine. Whether you're a hobbyist or a professional, this chapter will equip you with the knowledge and skills to keep your laser engraver in peak condition, ensuring it continues to deliver stunning results for years to come. So, let's dive into the world of laser engraver maintenance and unlock the secrets to a long-lasting and productive partnership!

9.1 Regular Cleaning and Maintenance

Regular cleaning and maintenance are the cornerstones of keeping your laser engraver in optimal condition. Just like any other machine, your engraver requires routine care to ensure smooth operation, accurate results, and a long lifespan.

- **Daily Cleaning:** After each use, clean the work area and remove any debris or residue from

the bed, mirrors, and lenses. Use a soft, lint-free cloth and a specialized cleaning solution recommended by the manufacturer. Avoid using harsh chemicals or abrasive materials that could damage the delicate surfaces.
- **Weekly Cleaning:** Once a week, perform a more thorough cleaning of the laser engraver. This includes cleaning the exhaust fan, filter, and any other accessible components. Inspect the wiring for any damage or wear and tear, and tighten any loose screws or connections.
- **Monthly Cleaning:** Once a month, conduct a deep cleaning of the laser engraver. This involves disassembling some parts, cleaning them thoroughly, and reassembling them according to the manufacturer's instructions. Pay close attention to the laser head, mirrors, and lenses, as these are critical for engraving and cutting accuracy.
- **Lubrication:** Some laser engravers have moving parts that require periodic lubrication. Refer to the user manual for instructions on how to lubricate these parts properly. Use a lubricant recommended by the manufacturer and avoid over-lubricating, as this can attract dust and debris.

By establishing a regular cleaning and maintenance routine, you can prevent dust and debris buildup, minimize wear and tear, and ensure that your laser engraver continues to perform at its best.

9.2 Replacing Parts and Consumables

Like any machine, laser engravers have parts and consumables that wear out over time and need to be replaced. These include:

- **Laser Tube:** The laser tube is the heart of the laser engraver, generating the powerful beam of light used for engraving and cutting. Laser tubes have a limited lifespan and will eventually need to be replaced. The lifespan of a laser tube depends on its quality, usage frequency, and maintenance practices.
- **Mirrors and Lenses:** Mirrors and lenses are used to direct and focus the laser beam. They can become dirty, scratched, or misaligned over time, affecting the engraver's performance. Regular cleaning can help extend their lifespan, but eventually, they will need to be replaced.
- **Focus Lens:** The focus lens is a critical component that determines the spot size and intensity of the laser beam. It can become contaminated or damaged over time, leading to blurry or uneven engravings and cuts.
- **Air Assist Nozzle:** The air assist nozzle directs a stream of air onto the cutting area to remove debris and cool the material. It can become clogged with dust or debris, affecting its performance.
- **Exhaust Fan and Filter:** The exhaust fan and filter are responsible for removing smoke and fumes generated during engraving and cutting.

The filter can become clogged over time, reducing its effectiveness and potentially harming the engraver.
- **Belts and Pulleys:** Some laser engravers use belts and pulleys to move the laser head and other components. These can wear out over time and may need to be replaced to ensure smooth and accurate movement.

Refer to the user manual for instructions on how to replace these parts and consumables. It's important to use genuine parts from the manufacturer to ensure compatibility and optimal performance.

9.3 Troubleshooting Common Problems

Despite regular maintenance, you may encounter some common problems with your laser engraver. Here are some troubleshooting tips for addressing typical issues:

- **Laser Beam Not Firing:** Check the power connections, laser tube connections, and water cooling system (if applicable). Ensure that the software is sending the correct commands to the engraver. If the problem persists, consult the user manual or contact the manufacturer's support.
- **Blurry or Uneven Engravings:** This can be caused by a dirty lens or mirror, improper focus, or vibrations during the engraving process. Clean the lens or mirror with a soft

cloth and specialized cleaning solution. Check for loose parts or vibrations in the engraver and tighten or adjust them as needed.
- **Inconsistent Engraving Depth:** This can be caused by variations in material thickness, inconsistent power delivery from the laser, or a worn-out laser tube. Check the material for thickness variations and ensure that it is level on the bed. If the power delivery is inconsistent or the laser tube is worn out, you may need to replace the laser tube or contact the manufacturer for support.
- **Skipped Lines or Areas:** This can occur if the speed setting is too high for the laser to accurately follow the design. Reduce the speed setting to allow the laser more time to engrave each line or area.
- **Excessive Noise or Vibration:** This can be caused by loose parts, worn-out bearings, or other mechanical issues. Tighten any loose screws or bolts, replace worn-out bearings, and check for any obstructions in the engraver's path.
- **Error Messages:** If you see an error message on the engraver's display or in the software, consult the user manual or contact the manufacturer's support for assistance.

By following these troubleshooting tips, you can often resolve common problems and keep your laser engraver running smoothly. However, if you're unable to resolve the issue on your own, don't hesitate to seek help from online forums, user groups, or the manufacturer's support team.

9.4 Extending the Lifespan of Your Engraver

By taking good care of your laser engraver and following the manufacturer's recommendations, you can significantly extend its lifespan and ensure optimal performance for years to come. Here are some additional tips for maximizing the longevity of your engraver:

- **Avoid Overheating:** Laser tubes and other components can be damaged by overheating. Ensure that the engraver is adequately ventilated and that the cooling system (if applicable) is functioning properly.
- **Use Quality Materials:** Using high-quality materials for engraving and cutting can reduce wear and tear on the laser tube and other components. Avoid engraving or cutting materials that are known to be abrasive or corrosive.
- **Store Properly:** When not in use, store the laser engraver in a clean, dry, and dust-free environment. Cover it with a protective cloth or case to prevent damage from dust and moisture.
- **Regularly Check for Updates:** Manufacturers often release software updates that improve performance, fix bugs, and add new features.

Keep your software up-to-date to ensure optimal compatibility and functionality.
- **Professional Servicing:** Consider having your laser engraver professionally serviced once a year or according to the manufacturer's recommendations. This can help identify and address potential problems before they cause serious damage.

By following these tips and taking a proactive approach to maintenance, you can ensure that your laser engraver continues to be a reliable and valuable tool for your creative endeavors.

9.5 Safety Checks and Inspections

Regular safety checks and inspections are essential for ensuring the safe operation of your laser engraver. Here are some key areas to check:

- **Laser Safety Interlocks:** Most laser engravers have safety interlocks that prevent the laser from firing if the enclosure is open or if other safety conditions are not met. Check these interlocks regularly to ensure they are functioning properly.
- **Emergency Stop Button:** Ensure that the emergency stop button is easily accessible and functioning correctly. Test it periodically to ensure that it stops the engraver immediately.
- **Ventilation:** Check the exhaust fan and filter to ensure they are clean and functioning

properly. Replace the filter if it's clogged or damaged.
- **Wiring and Connections:** Inspect all wiring and connections for any signs of damage or wear and tear. Tighten any loose connections and replace any damaged wires or cables.
- **Grounding:** Ensure that the laser engraver is properly grounded to prevent electrical shocks.

If you notice any problems during your safety checks or inspections, address them immediately. Don't use the engraver until the problems have been resolved.

9.6 Calibrating and Aligning the Laser

Over time, the laser beam's alignment and focus may drift slightly, affecting the accuracy of your engravings and cuts. Regular calibration and alignment can help ensure consistent and precise results.

- **Calibration:** Calibration involves adjusting the laser's power, speed, and focus settings to match the specific materials you're using. Refer to the user manual for instructions on how to calibrate your laser engraver.
- **Alignment:** Alignment involves adjusting the mirrors and lenses to ensure that the laser beam is properly focused and centered on the material. This is typically done using specialized tools and procedures provided by the manufacturer.

It's recommended to calibrate and align your laser engraver every few months or whenever you notice a decrease in engraving or cutting quality.

9.7 Professional Servicing and Repairs

While regular maintenance and cleaning can help prevent most problems, there may be times when your laser engraver requires professional servicing or repairs. If you encounter a problem that you can't resolve on your own, or if you notice any unusual noises, smells, or performance issues, it's best to contact the manufacturer or a qualified technician for assistance.

Professional servicing can include:

- Cleaning and maintenance of all components
- Replacement of worn-out parts or consumables
- Calibration and alignment
- Firmware updates (if applicable)
- Troubleshooting and repair of complex issues

Regular professional servicing can help ensure that your laser engraver is always in top condition and operating safely. It can also help identify potential problems before they cause major damage, saving you time and money in the long run.

Maintaining your laser engraver is an ongoing process that requires diligence and care. By following the guidelines outlined in this chapter, you can ensure that your engraver remains a reliable and high-performing tool for years to come.

Remember, a well-maintained laser engraver not only produces better results but also operates more safely and efficiently. Make regular cleaning and maintenance a priority, and don't hesitate to seek professional help when needed.

By investing in the care of your laser engraver, you're investing in your creativity and your ability to bring your ideas to life.

CHAPTER 10

ADVANCED SOFTWARE AND DESIGN

Welcome to the realm of design wizardry! In this chapter, we'll transcend the basics of laser engraving and delve into the captivating world of advanced software and design techniques. Prepare to elevate your skills, unleash your creativity, and unlock the full potential of your laser engraver. We'll embark on a journey through the intricacies of complex designs, harness the precision of CAD software, and even venture into the realm of 3D modeling. By the end of this chapter, you'll be equipped with a powerful arsenal of tools and techniques to create awe-inspiring laser engravings that push the boundaries of artistry and innovation. So, let's dive in and discover the boundless possibilities that await you!

10.1 Exploring Advanced Design Software

While basic design software like Inkscape and CorelDRAW offer a solid foundation for laser engraving, exploring advanced software can open up a whole new world of creative possibilities. These powerful tools provide an extensive array of features and functionalities that cater to professional designers and laser engraving enthusiasts alike.

- **Adobe Illustrator:** This industry-standard vector graphics editor is renowned for its robust toolset, intuitive interface, and versatility. With Illustrator, you can create intricate designs, manipulate complex shapes, and seamlessly integrate text and images. Its extensive library of brushes, patterns, and effects provides endless options for customization and personalization.
- **Affinity Designer:** A rising star in the design software world, Affinity Designer offers a compelling alternative to Adobe Illustrator. With its sleek interface, powerful features, and affordable price point, Affinity Designer is gaining popularity among designers and laser engraving enthusiasts. It boasts an impressive array of tools for creating vector illustrations, manipulating images, and designing layouts.
- **CorelDRAW Graphics Suite:** This comprehensive suite of design software includes CorelDRAW for vector illustration, Corel PHOTO-PAINT for image editing, and Corel Font Manager for organizing and managing fonts. CorelDRAW's extensive toolset, intuitive interface, and wide range of features make it a versatile choice for graphic design, laser engraving, and other creative projects.
- **Inkscape (Advanced Features):** While primarily known as a beginner-friendly software, Inkscape also offers a wealth of advanced features for those who want to take their designs to the next level. These features

include node editing, path operations, live path effects, and extensions for specialized tasks like 3D modeling and G-code generation.
- **Other Specialized Software:** Depending on your specific needs and preferences, you might also consider exploring other specialized software options. For example, if you're interested in creating intricate patterns or generating complex toolpaths, you might look into software like Autodesk Fusion 360 or Vectric Aspire.

Remember, choosing the right design software is a personal decision. Experiment with different options to find one that aligns with your workflow, budget, and creative vision.

10.2 Creating Complex Designs

With the power of advanced design software, you can create intricate and visually stunning designs that were previously unimaginable. Here are some techniques to elevate your designs to the next level:

- **Layering:** By layering multiple elements within your design, you can create depth, dimension, and visual interest. Experiment with different opacities, blending modes, and layer effects to achieve the desired look.
- **Gradients and Blends:** Gradients and blends can add a touch of realism and complexity to your designs. Use them to create smooth transitions between colors, simulate lighting

effects, or add subtle texture to your engravings.
- **Clipping Masks:** Clipping masks allow you to confine an image or design within a specific shape, creating interesting visual effects and adding a touch of creativity to your engravings.
- **Compound Paths:** Compound paths combine multiple shapes into a single entity, allowing you to manipulate and transform them as a unit. This technique is useful for creating intricate patterns, logos, and complex designs.
- **Custom Brushes and Patterns:** Most advanced design software allows you to create your own custom brushes and patterns. This gives you endless possibilities for adding unique textures, patterns, and artistic flair to your engravings.
- **Text Effects:** Experiment with different text effects like warping, outlining, and drop shadows to create eye-catching text elements that enhance the overall visual appeal of your designs.
- **Symbol Libraries:** Many design software programs come with built-in symbol libraries that contain a wide range of pre-designed elements like icons, shapes, and illustrations. You can also create your own symbol libraries to save time and ensure consistency across your projects.

By mastering these advanced design techniques, you can create complex and visually stunning designs that will leave a lasting impression.

10.3 Using CAD Software for Precision

Computer-Aided Design (CAD) software is a powerful tool for creating precise and accurate designs, especially for engineering, architecture, and manufacturing applications. While CAD software may seem intimidating at first, it offers unparalleled precision and control over your designs, making it a valuable asset for laser engraving projects that require high levels of accuracy and detail.

- **Precision Drawing Tools:** CAD software provides a vast array of precision drawing tools that allow you to create complex shapes, curves, and angles with exceptional accuracy. You can specify exact dimensions, angles, and tolerances to ensure that your designs are perfectly aligned and fit together seamlessly.
- **3D Modeling:** Some CAD software programs also offer 3D modeling capabilities, allowing you to create three-dimensional representations of your designs. This can be helpful for visualizing how your design will look when engraved and for creating 3D models for 3D printing or other manufacturing processes.
- **Assembly and Simulation:** CAD software often includes features for assembling multiple parts into a single design and simulating how

they will interact with each other. This can be useful for ensuring that your design is functional and that all the parts fit together properly.
- **Technical Drawings:** CAD software allows you to create detailed technical drawings that include dimensions, tolerances, and other specifications. These drawings are essential for communicating your design intent to manufacturers or other stakeholders.

By incorporating CAD software into your workflow, you can achieve a level of precision and accuracy that is simply not possible with traditional design tools. This is especially important for projects that require tight tolerances or intricate details.

10.4 Incorporating 3D Modeling

3D modeling is another powerful tool that can be used in conjunction with laser engraving. By creating a 3D model of your design, you can visualize how it will look when engraved and make any necessary adjustments before committing to the final engraving.

There are several ways to incorporate 3D modeling into your laser engraving workflow:

- **Creating 3D Models from Scratch:** If you have experience with 3D modeling software, you can create 3D models of your designs from scratch. This gives you complete control over

the shape, dimensions, and details of the model.
- **Importing 3D Models:** If you don't have 3D modeling skills, you can import 3D models from online libraries or create them using online tools like Tinkercad or SketchUp. These tools are relatively easy to use and allow you to create simple 3D models without requiring extensive training.
- **Converting 2D Designs to 3D:** Some laser engraving software programs offer tools for converting 2D designs into 3D models. This can be a quick and easy way to add depth and dimension to your engravings.

Once you have a 3D model of your design, you can use laser engraving software to generate a toolpath that the laser engraver will follow to create the 3D engraving. This process typically involves slicing the 3D model into layers and then engraving each layer onto the material.

10.5 Generating G-Code for Engraving

G-code is a programming language used to control CNC (Computer Numerical Control) machines, including laser engravers. G-code files contain instructions for the machine, such as where to move the laser head, how fast to move it, and at what power level to fire the laser.

Most laser engraving software programs can generate G-code files based on your design. This allows you to send your designs directly to the laser engraver without having to manually program the machine.

To generate G-code for your design, you'll typically need to specify the following parameters in your laser engraving software:

- **Material Type:** The type of material you'll be engraving, as this will affect the laser settings.
- **Engraving Depth:** The desired depth of the engraving.
- **Laser Power:** The power level of the laser beam.
- **Engraving Speed:** The speed at which the laser beam moves across the material.
- **DPI:** The resolution of the engraving.

Once you've specified these parameters, the software will generate a G-code file that you can save and send to your laser engraver.

10.6 Optimizing Designs for Efficiency

When designing for laser engraving, it's important to consider not only the aesthetic appeal of your design but also its efficiency. An efficient design can save you time and money by minimizing the amount of material used and reducing the engraving time.

Here are some tips for optimizing your designs for efficiency:

- **Nesting:** Nesting involves arranging multiple parts of a design in a way that minimizes wasted space on the material. This can be done manually or using specialized nesting software.
- **Common Line Cutting:** Common line cutting involves identifying lines that are shared by multiple parts of a design and cutting them only once. This can save time and reduce wear and tear on the laser engraver.
- **Minimizing Travel Distance:** Arrange the elements of your design in a way that minimizes the distance the laser head needs to travel between cuts. This can reduce engraving time and improve overall efficiency.
- **Using Vector Images:** Vector images are generally more efficient to engrave than raster images because they require less data to represent the design.

10.7 Troubleshooting Software Issues

Despite the power and sophistication of advanced design software, you may encounter occasional glitches or errors. Here are some troubleshooting tips for common software issues:

- **Software Crashes:** If your design software crashes unexpectedly, try restarting your computer and the software. If the problem persists, check for software updates or reinstall the software.

- **File Corruption:** If you're unable to open or save a design file, it may be corrupted. Try opening the file in a different software program or using a file repair tool. If the file is severely corrupted, you may need to recreate the design from scratch.
- **Compatibility Issues:** If you're having trouble importing or exporting files, ensure that the file format is compatible with your software. Try converting the file to a different format or updating your software to the latest version.
- **Missing Fonts:** If your design uses a font that is not installed on your computer, it may not display correctly. Install the missing font or substitute it with a similar font that is available.
- **Plugin Errors:** If you're using plugins or extensions, they may conflict with your software or cause errors. Try disabling the plugins or extensions one by one to see if they are causing the problem. If you identify a problematic plugin or extension, try updating it or removing it altogether.
- **Rendering Issues:** If your design is slow to render or appears distorted, try simplifying the design or reducing the number of layers. You can also try adjusting the rendering settings in your software.
- **Hardware Issues:** If you're experiencing persistent software issues, there may be a problem with your computer's hardware, such as insufficient memory or a faulty graphics

card. Try upgrading your hardware or consult a computer technician for assistance.

If you're unable to resolve the software issue on your own, consult the software's documentation, online forums, or the software provider's support for further assistance.

Advanced software and design techniques empower you to create laser engravings that are not only visually stunning but also precise, accurate, and efficient. By mastering these tools and techniques, you can unlock a world of creative possibilities and elevate your laser engraving skills to new heights.

Remember, the key to success is continuous learning and experimentation. Explore different software options, practice advanced design techniques, and embrace the power of 3D modeling and CAD software to create truly unique and personalized projects. With dedication and passion, you can unleash your creativity and transform your ideas into breathtaking laser engravings that inspire and amaze.

CHAPTER 11

BUSINESS AND ENTREPRENEURSHIP WITH LASER ENGRAVING

Welcome to the realm of entrepreneurial adventure! In this chapter, we'll shift gears from the artistic and technical aspects of laser engraving to explore the exciting world of business and entrepreneurship. If you've ever dreamed of turning your passion for laser engraving into a profitable venture, this chapter is your roadmap to success. We'll delve into the intricacies of starting a laser engraving business, from developing a solid business plan to marketing your services, pricing your products, managing orders and production, building a loyal customer base, and even expanding your operations. We'll also touch upon the legal and regulatory considerations that are essential for running a successful business. So, if you're ready to take the leap and transform your creative skills into a thriving enterprise, let's dive in and unlock the secrets to building a successful laser engraving business!

11.1 Starting a Laser Engraving Business

Embarking on a new business venture is both exciting and challenging. To ensure your laser engraving business thrives, it's crucial to lay a solid foundation. Here's a step-by-step guide to help you get started:

1. **Develop a Business Plan:** A comprehensive business plan is your roadmap to success. It outlines your vision, mission, target market, marketing strategies, financial projections, and operational plans. Take the time to research your market, identify your competitors, and define your unique selling proposition (USP).
2. **Choose a Business Structure:** Decide on the legal structure of your business, whether it's a sole proprietorship, partnership, limited liability company (LLC), or corporation. Each structure has its own advantages and disadvantages regarding liability, taxation, and ownership. Consult with a legal professional or accountant to determine the best option for your situation.
3. **Secure Funding:** Determine how you'll finance your business. This could include personal savings, loans from family and friends, small business loans, or crowd funding. Research different funding options and choose the one that best suits your needs and financial situation.
4. **Obtain Licenses and Permits:** Depending on your location and business activities, you may

need to obtain various licenses and permits to operate legally. Contact your local government agencies to determine the specific requirements for your business.
5. **Purchase or Lease Equipment:** Invest in a high-quality laser engraver that meets your business needs. Consider factors like engraving area, power, speed, supported materials, and software compatibility. If you're starting on a tight budget, you might consider leasing equipment initially.
6. **Set Up Your Workspace:** Create a dedicated workspace for your laser engraver. Ensure that it's well-ventilated, has adequate lighting, and meets safety standards. Consider investing in additional equipment like fume extractors, ventilation systems, and safety gear to create a safe and comfortable working environment.
7. **Develop Your Skills:** Continuously refine your laser engraving skills and stay up-to-date on the latest techniques and trends. Take online courses, attend workshops, and experiment with different materials and designs to expand your expertise.
8. **Build a Portfolio:** Create a portfolio of your best work to showcase your skills and attract potential clients. Include a variety of projects that demonstrate your versatility and creativity.
9. **Network and Build Relationships:** Connect with other professionals in the industry, attend trade shows, and join online communities to network and build relationships. Networking

can open doors to new opportunities, collaborations, and referrals.
10. **Develop a Pricing Strategy:** Determine your pricing structure for products and services. Consider factors like material costs, labor, overhead expenses, and desired profit margins. Research your competitors' pricing to ensure your rates are competitive.

By following these steps and investing time and effort into building a solid foundation, you'll be well on your way to launching a successful laser engraving business.

11.2 Marketing and Promoting Your Services

Once your business is up and running, it's time to spread the word and attract customers. Marketing and promotion are essential for generating awareness, building your brand, and driving sales. Here are some effective strategies to consider:

1. **Create a Professional Website:** A well-designed website is your online storefront. It should showcase your portfolio, services, pricing, contact information, and testimonials from satisfied customers. Invest in professional web design and ensure that your website is mobile-friendly and easy to navigate.
2. **Social Media Marketing:** Leverage social media platforms like Facebook, Instagram, Pinterest, and Twitter to promote your services

and connect with potential customers. Share high-quality photos and videos of your work, offer special promotions, and engage with your followers to build relationships.
3. **Content Marketing:** Create informative and engaging content like blog posts, articles, tutorials, and videos to share your expertise and attract potential customers. Share this content on your website and social media channels to drive traffic and build brand awareness.
4. **Email Marketing:** Build an email list of potential and existing customers and send them regular newsletters highlighting your latest projects, promotions, and industry news. Email marketing is a cost-effective way to nurture leads and drive repeat business.
5. **Local SEO:** Optimize your website and online listings for local search engines. This will help potential customers in your area find you when they search for laser engraving services. Claim your business listings on Google My Business, Yelp, and other directories.
6. **Paid Advertising:** Consider investing in paid advertising platforms like Google Ads and social media ads to reach a wider audience and target specific demographics.
7. **Partnerships and Collaborations:** Partner with complementary businesses like gift shops, wedding planners, or eventorganizers to cross-promote your services and reach new audiences.

8. **Attend Trade Shows and Events:** Participate in local craft fairs, trade shows, and industry events to showcase your work, network with potential clients and partners, and gain exposure for your brand.

By implementing a multi-faceted marketing strategy, you can effectively reach your target audience and generate leads for your laser engraving business.

11.3 Pricing Your Products and Services

Pricing your products and services is a critical aspect of running a profitable business. Here are some factors to consider when developing your pricing strategy:

1. **Cost of Goods Sold (COGS):** Calculate the cost of materials, labor, and any other direct expenses involved in creating your products or providing your services. This will give you a baseline for setting your prices.
2. **Overhead Expenses:** Consider your fixed costs, such as rent, utilities, insurance, software subscriptions, and marketing expenses. Factor these costs into your pricing to ensure you're covering your expenses.
3. **Profit Margin:** Determine your desired profit margin. This is the percentage of each sale that you keep as profit after deducting your expenses. Aim for a profit margin that allows you to reinvest in your business, pay yourself a fair salary, and grow your operations.

4. **Market Research:** Research your competitors' pricing to ensure your rates are competitive. However, don't undercut your prices too much, as this can devalue your services and make it difficult to sustain your business.
5. **Value-Based Pricing:** Consider the value you're providing to your customers. If you're offering unique or high-quality products or services, you can charge a premium price.
6. **Pricing Models:** Choose a pricing model that suits your business and your customers. You can charge per hour, per project, or offer package deals for multiple items or services.
7. **Discounts and Promotions:** Offer discounts or promotions to attract new customers and incentivize repeat business. This could include seasonal discounts, referral programs, or bulk order discounts.

By carefully considering these factors and developing a clear pricing strategy, you can ensure that your laser engraving business is profitable and sustainable.

11.4 Managing Orders and Production

Efficient order management and production are essential for delivering high-quality products and services on time and within budget. Here are some tips for streamlining your operations:

1. **Implement an Order Management System:** Use a software program or online platform to manage customer orders, track inventory, and

schedule production. This will help you stay organized and ensure that orders are fulfilled accurately and efficiently.
2. **Streamline Your Workflow:** Develop a clear and efficient workflow for each type of product or service you offer. This could involve creating templates, checklists, or standard operating procedures (SOPs) to ensure consistency and minimize errors.
3. **Optimize Production:** Look for ways to optimize your production process to reduce costs and improve efficiency. This could involve investing in automation tools, streamlining your design process, or outsourcing certain tasks.
4. **Maintain Inventory:** Keep track of your inventory levels and reorder materials as needed to avoid running out of stock and causing delays for your customers.
5. **Quality Control:** Implement a rigorous quality control process to ensure that every product or service you deliver meets your high standards. Inspect each item before shipping or delivering it to the customer.
6. **Communication:** Communicate clearly and regularly with your customers throughout the order fulfillment process. Provide updates on their orders, answer any questions they may have, and address any concerns promptly.

By implementing efficient order management and production practices, you can ensure that your customers are satisfied and that your business runs smoothly.

11.5 Building a Customer Base

Building a loyal customer base is essential for the long-term success of your laser engraving business. Here are some strategies to attract and retain customers:

1. **Provide Excellent Customer Service:** Go above and beyond to meet your customers' needs and exceed their expectations. Respond to inquiries promptly, offer helpful advice, and resolve any issues quickly and professionally.
2. **Offer High-Quality Products and Services:** Deliver exceptional quality in every product and service you offer. Use high-quality materials, pay attention to detail, and strive for perfection in every engraving or cutting project.
3. **Personalize the Experience:** Offer personalized touches that make your customers feel valued and appreciated. This could include handwritten thank-you notes, custom packaging, or special offers for returning customers.
4. **Build Relationships:** Get to know your customers and their needs. Ask for feedback, offer suggestions, and engage with them on social media to build lasting relationships.
5. **Loyalty Programs:** Implement a loyalty program to reward repeat customers. This

could include discounts, exclusive offers, or early access to new products or services.
6. **Referral Programs:** Encourage your customers to refer their friends and family to your business by offering incentives like discounts or free gifts.
7. **Community Involvement:** Get involved in your local community by sponsoring events, donating to charities, or volunteering your time. This can help raise awareness of your business and build goodwill among potential customers.

By focusing on customer satisfaction, building relationships, and offering exceptional value, you can cultivate a loyal customer base that will support your business for years to come.

11.6 Expanding Your Business

Once your laser engraving business is established and profitable, you may consider expanding your operations. Here are some ways to grow your business:

1. **Offer New Products or Services:** Expand your product or service offerings to attract new customers and cater to a wider range of needs. This could involve adding new engraving materials, offering custom design services, or expanding into related fields like 3D printing or CNC machining.

2. **Increase Production Capacity:** Invest in additional equipment or hire more staff to increase your production capacity and fulfill larger orders. This will allow you to take on more projects and grow your revenue.
3. **Expand Your Market Reach:** Explore new markets or demographics to expand your customer base. This could involve targeting different industries, geographic regions, or online platforms.
4. **Franchise or License Your Brand:** If you have a successful business model, you could consider franchising or licensing your brand to allow others to open their own laser engraving businesses under your name and guidance.
5. **Acquire or Merge with Another Business:** Acquiring or merging with another business can be a quick way to expand your operations, gain access to new markets, and increase your customer base.

Expanding your business requires careful planning and execution. Before taking any major steps, conduct thorough market research, assess your financial resources, and develop a solid expansion strategy.

11.7 Legal and Regulatory Considerations

Running a laser engraving business involves complying with various legal and regulatory requirements. These requirements can vary depending on your location and the nature of your business activities. Here are some key considerations:

1. **Business Registration:** Register your business with the appropriate government agencies and obtain any necessary licenses and permits.
2. **Tax Compliance:** Comply with all applicable tax laws, including income tax, sales tax, and payroll tax. Consult with a tax professional to ensure you're meeting all your obligations.
3. **Product Safety:** If you're selling products, ensure they comply with relevant safety standards and regulations. This may involve obtaining product certifications, labeling requirements, and safety testing.
4. **Intellectual Property:** Protect your intellectual property, such as your business name, logo, and designs, by registering trademarks and copyrights.
5. **Liability Insurance:** Obtain liability insurance to protect your business from claims of injury or damage caused by your products or services.
6. **Employee Safety:** If you have employees, ensure that you're providing a safe and healthy working environment and complying with all applicable labor laws.
7. **Environmental Regulations:** Laser engraving can produce fumes and particulate matter. Comply with all environmental regulations regarding air quality and waste disposal.

By understanding and complying with legal and regulatory requirements, you can protect your business from legal issues and ensure its long-term success.

Building a successful laser engraving business requires a combination of creativity, technical skills, business acumen, and a passion for entrepreneurship. By developing a solid business plan, marketing your services effectively, pricing your products strategically, managing orders and production efficiently, building a loyal customer base, and complying with legal and regulatory requirements, you can turn your passion for laser engraving into a thriving and fulfilling enterprise.

Remember, the journey of entrepreneurship is filled with challenges and rewards. Embrace the challenges as opportunities for growth, learn from your mistakes, and celebrate your successes. With dedication, perseverance, and a willingness to adapt and evolve, you can achieve your dreams and build a laser engraving business that makes a lasting impact.

CHAPTER 12

THE FUTURE OF LASER ENGRAVING

Welcome to the final chapter of our laser engraving journey! In this chapter, we'll take a step into the future and explore the exciting possibilities that lie ahead for this dynamic technology. From groundbreaking advancements in laser technology and innovative applications to the integration of artificial intelligence and a growing focus on sustainability, the future of laser engraving is brimming with potential. We'll delve into emerging trends, new materials and applications, advancements in software, the environmental impact, the role of AI, career opportunities in the field, and valuable resources for further learning. Whether you're a seasoned engraver or a budding enthusiast, this chapter will ignite your imagination and inspire you to embrace the boundless possibilities that await in the ever-evolving world of laser engraving.

12.1 Emerging Trends and Technologies

The laser engraving landscape is constantly evolving, with new trends and technologies emerging at a rapid pace. Here are some key developments shaping the future of laser engraving:

- **Advanced Laser Sources:** Laser technology continues to advance, with the development of more powerful, efficient, and versatile laser sources. Diode lasers, for instance, are becoming increasingly popular due to their compact size, affordability, and ability to engrave various materials. Additionally, ultrafast lasers, such as femtosecond lasers, are gaining traction in precision micromachining applications due to their ultra-short pulse durations and minimal heat-affected zones.
- **Integration of Automation and Robotics:** Automation and robotics are becoming increasingly integrated into laser engraving processes, enhancing efficiency, precision, and productivity. Robotic arms equipped with laser heads can perform complex engraving tasks with unmatched speed and accuracy, freeing up human operators for more creative and strategic roles.
- **Hybrid Laser Systems:** Hybrid laser systems that combine different laser sources, such as CO_2 and fiber lasers, are gaining popularity due to their versatility and ability to engrave a wider range of materials. These systems offer the benefits of both laser types, allowing for greater flexibility and customization in engraving applications.
- **Portable and Handheld Laser Engravers:** The advent of portable and handheld laser engravers is revolutionizing the accessibility and convenience of laser engraving. These compact devices allow for on-the-go

engraving, making it easier than ever to personalize items, create custom designs, and mark objects in various settings.

- **Customization and Personalization:** The demand for customized and personalized products continues to grow, driving innovation in laser engraving technology. Laser engravers are increasingly used to create unique gifts, personalized items, and custom-made products that cater to individual preferences and styles.

These emerging trends and technologies are transforming the laser engraving landscape, opening up new possibilities for creativity, productivity, and business growth. As technology continues to advance, we can expect even more exciting developments in the years to come.

12.2 New Materials and Applications

The versatility of laser engraving is constantly expanding with the discovery of new materials and applications. Here are some exciting developments in this area:

- **Biodegradable Materials:** As sustainability becomes a top priority, laser engraving is increasingly being used on biodegradable materials like wood, bamboo, and plant-based plastics. This allows for the creation of eco-friendly products that minimize environmental impact.

- **Advanced Polymers:** The development of advanced polymers with unique properties, such as heat resistance, flexibility, and transparency, is opening up new avenues for laser engraving. These materials can be used to create innovative products like flexible electronics, wearable devices, and custom medical implants.
- **Textiles and Fabrics:** Laser engraving is finding new applications in the textile and fashion industry. It can be used to create intricate patterns, logos, and designs on fabrics, adding a unique touch to clothing, accessories, and home decor items.
- **Food and Beverages:** Laser engraving is even being used to personalize food and beverage items like cakes, cookies, fruits, and bottles. This adds a touch of customization and novelty to culinary creations.
- **Medical Applications:** Laser engraving is increasingly used in the medical field for marking surgical instruments, implants, and prosthetics. This ensures traceability, improves patient safety, and aids in inventory management.

These are just a few examples of the expanding range of materials and applications for laser engraving. As researchers and innovators continue to push the boundaries, we can expect even more exciting developments in this field.

12.3 Innovations in Laser Engraving Software

Software plays a crucial role in laser engraving, enabling users to design, control, and optimize the engraving process. Recent innovations in laser engraving software are making the technology more accessible, user-friendly, and powerful than ever before.

- **User-Friendly Interfaces:** Laser engraving software is becoming increasingly intuitive and user-friendly, with streamlined workflows and simplified controls. This makes it easier for beginners and non-technical users to create and execute complex engraving projects.
- **Cloud-Based Software:** Cloud-based laser engraving software allows users to access their designs and control their engravers from anywhere with an internet connection. This offers greater flexibility and collaboration opportunities, especially for businesses and teams working remotely.
- **Artificial Intelligence (AI) Integration:** AI is being integrated into laser engraving software to automate tasks, optimize settings, and enhance design capabilities. For example, AI algorithms can analyze images and automatically generate optimized engraving paths, saving time and improving efficiency.
- **Simulation and Preview Tools:** Advanced software now offers simulation and preview tools that allow users to visualize how their

designs will look on different materials and with different settings before actually engraving them. This helps to minimize errors and waste, saving time and money.
- **Integration with 3D Modeling Software:** Laser engraving software is increasingly integrated with 3D modeling software, allowing for seamless transfer of 3D designs and generation of optimized engraving paths. This simplifies the workflow for creating complex 3D engravings.

These innovations in laser engraving software are empowering users to achieve greater precision, efficiency, and creativity in their projects. As software continues to evolve, we can expect even more powerful and user-friendly tools to emerge, further democratizing the field of laser engraving.

12.4 Environmental Impact and Sustainability

As with any technology, laser engraving has an environmental impact. The production and use of laser engravers consume energy and resources, and the engraving process itself can generate waste and emissions. However, there are several ways to mitigate the environmental impact and promote sustainability in laser engraving:

- **Energy-Efficient Laser Engravers:** Choose laser engravers that are designed for energy efficiency. Look for models with features like

sleep mode, automatic power-off, and energy-saving settings.
- **Sustainable Materials:** Whenever possible, use sustainable materials for your engraving projects. This could include recycled materials, biodegradable materials, or materials sourced from responsibly managed forests.
- **Waste Reduction:** Minimize waste by optimizing your designs, using scrap materials for testing, and reusing or recycling materials whenever possible.
- **Proper Disposal:** Dispose of waste materials, such as used filters and laser tubes, responsibly according to local regulations.
- **Energy Conservation:** Turn off the laser engraver when not in use, and unplug it when not in use for extended periods.

By taking these steps, you can reduce the environmental impact of your laser engraving activities and contribute to a more sustainable future.

12.5 The Role of AI in Laser Engraving

Artificial intelligence (AI) is rapidly transforming various industries, and laser engraving is no exception. AI is being integrated into laser engraving software and hardware to automate tasks, optimize processes, and enhance creativity. Here are some ways AI is shaping the future of laser engraving:

- **Automated Design Generation:** AI algorithms can analyze images and generate optimized

engraving paths, saving time and effort for designers. This can be particularly useful for complex or repetitive designs.
- **Adaptive Laser Control:** AI can be used to monitor and adjust laser parameters in real-time based on the material being engraved and the desired outcome. This can improve engraving quality, speed, and consistency.
- **Predictive Maintenance:** AI can analyze data from laser engravers to predict when maintenance or repairs may be needed. This can help prevent breakdowns and extend the lifespan of the equipment.
- **Quality Control:** AI algorithms can be used to inspect engraved products for defects or inconsistencies, ensuring that only high-quality products are shipped to customers.
- **Personalized Recommendations:** AI can be used to analyze customer data and preferences to provide personalized recommendations for products and designs. This can enhance the customer experience and drive sales.

The integration of AI in laser engraving is still in its early stages, but it has the potential to revolutionize the industry. As AI technology continues to advance, we can expect even more innovative applications that will enhance productivity, creativity, and efficiency in laser engraving.

12.6 Career Opportunities in the Field

The field of laser engraving offers a wide range of career opportunities for individuals with diverse skills and interests. Here are some potential career paths:

- **Laser Engraving Technician:** Laser engraving technicians operate and maintain laser engraving machines. They are responsible for setting up the machines, loading and unloading materials, and ensuring that the engravings meet quality standards.
- **Laser Engraving Designer:** Laser engraving designers create custom designs for a variety of applications, such as personalized gifts, signage, awards, and industrial markings. They use design software to create intricate and visually appealing designs that translate well to laser engraving.
- **Laser Engraving Engineer:** Laser engraving engineers develop and improve laser engraving technologies. They research new materials, design new laser systems, and optimize engraving processes for greater efficiency and precision.
- **Sales and Marketing Professionals:** Sales and marketing professionals promote laser engraving services and products to businesses and consumers. They develop marketing strategies, create promotional materials, and build relationships with clients.

- **Business Owners:** Entrepreneurs with a passion for laser engraving can start their own businesses, offering engraving services, selling custom products, or developing innovative laser engraving technologies.

The demand for skilled laser engraving professionals is expected to grow in the coming years, as the technology continues to expand into new markets.

Thank you for reading!!!

www.ingramcontent.com/pod-product-compliance
Lightning Source LLC
Chambersburg PA
CBHW071512220526
45472CB00003B/998